Builder's Guide to Contracting

Builder's Guide
to Contracting

John R. Zehner

Purchasing Agent and Assistant Secretary (retired)
Turner Construction Company

McGRAW-HILL BOOK COMPANY

New York   St. Louis   San Francisco   Düsseldorf   Johannesburg
Kuala Lumpur   London   Mexico   Montreal   New Delhi
Panama   Paris   São Paulo   Singapore
Sydney   Tokyo   Toronto

Library of Congress Cataloging in Publication Data

Zehner, John R
　　Builder's guide to contracting.

　　Bibliography: p.
　　1.　Building—Contracts and specifications.
I.　Title.
TH425.Z43　　　658′.99　　　74-19032
ISBN 0-07-072790-2

1234567890　KPKP　798765

The editors for this book were Jeremy Robinson and Margaret Lamb,
the designer was Naomi Auerbach, and the production supervisor
was George Oechsner.　It was set in Caledonia by Bi-Comp, Inc.

Printed and bound by The Kingsport Press.

To My Wife, Margaret

Contents

List of Illustrations

Preface

This guide is prepared for the construction man who is working in the field or in a contractor's or engineer's office. He may have just finished college or gained his education primarily through on-the-job experience. He may be operating as a general contractor with some jobs under his belt and looking for ideas on improving profits and obtaining more contracts. In any case, he has enthusiasm and enough experience to know that the construction business appeals to him and that he would like to grow with it.

General construction constitutes the largest industry in the United States. In the decade of the 1970s, it will employ over $3\frac{1}{4}$ million persons. It is an industry in which the majority of the building contractors and subcontractors function as individual operators. Figures of the Internal Revenue Service indicate that of 902,229 contractors business entities, 722,604 (81.2 percent) were sole proprietors, 50,562 (5.6 percent) were partnerships, and 129,063 (13.2 percent) were corporations. There is no single dominating company as in steel, automobiles, computers, or telephone communications. The ownership of construction companies is more completely held by active management than in any other business. Many subcontractors act directly as contractors in their agreements with owners or in subletting major portions of their subcontracts.

It is an easy business to get into but, as those in the trade know, there are complexities which soon come to the surface. The general contractor finds that his days are taken up with sales efforts, managing the business, community relations, labor and personnel problems, banking arrangements, and putting out the fires of troubles with which most contractors have to contend.

The business aspects of performing a construction contract constitute a major part of a general contractor's operations. Frequently, the financial success or failure of a job depends on the terms of the subcontracts and the pricing of the orders issued for the project. The book focuses on a phase of the business about which little has been written but which is one of the keys to success in the largest industry of the nation.

Technical competence and management responsibility is essential in operating a construction company and is usually obtained through field training. Training in buying subcontracts, on the other hand, is less likely to be obtained through field apprenticeship. There is less opportunity to gain experience through practice when the contractor's dollars are very much at stake and when decisions are made in the home office.

While the purchasing function involves vendor contacts and value analysis for construction materials, to a far greater degree in many companies the work provides for the procurement, evaluation, development, and management of subcontracts. It may be performed by the contractor himself or by an individual spending part of his time working in the field and part in the office, involved in estimating, job management, and purchasing. Larger companies usually have purchasing specialists and ultimately a full-scale purchasing department whose members function as part of a team for staffing the various jobs under the direction of a project manager or executive.

The general contractor's reputation and financial success as a builder will depend on many things including his ability to complete contracts on schedule with satisfactory workmanship. Closely related to his reputation will be the fairness of the methods used in negotiating with subcontractors and building material dealers.

These methods—the company's policies and procedures—need to be formalized to arrive at clearly defined, workable subcontracts that take into account the problems brought about by the progressive wage increases for shop and field labor, the far larger proportion of work handled through subcontractors, and the greater emphasis and need for scheduling to achieve beneficial occupancy or use of the building for the client, even though it may not be complete in all respects.

There is need for subcontracts to facilitate the efforts of those engaged in field operations. There is need also for agreements that will stand up in court if a subcontractor brings action against the contractor or if the subcontractor becomes bankrupt.

From trade publications such as *Contractor News* and *Engineering News-Record*, it is clear that contractors prepare and submit many hundreds of proposals each week to perform building projects of significant scope, exclusive of thousands of smaller jobs. Consistency in methods is desirable but, obviously, with such a great number of individuals involved, it is not surprising that there are radical differences in the procedures employed.

Although experience in preparing estimates is necessary, the general contractor is in business to build and not simply to secure training on dry runs. If ways can be determined in advance to arrive at lower subcontract costs, more favorable terms, or better cooperation from subcontractors, the chances increase that a greater proportion of the work for which he bids will be secured, overhead expenses will be decreased, and the opportunity for profits increased. In the home-building field, some of the methods used in subcontracting may result in lower costs with consequent easier sales to customers. While subcontracting is less extensive here, 40 percent of the contractors award subcontracts which amount to three-quarters or more of all costs for these companies.[1]

The risks are great and the rewards, occasionally, equally so. There are other situations where the contractor must assume responsibility for the shortcomings of others or of his own staff: bidding errors, financial difficulties, faulty material or workmanship, or inadequate plans and specifications, to mention only a few. For those who do not find the pot of gold at the end of the rainbow, the industry nevertheless remains as a last frontier of private enterprise where hundreds of thousands of contractors and subcontractors can and still do operate their own businesses successfully.

A guide of this nature obviously will not have all the answers. On the other hand, if it helps to chart a course for those individuals who sense the need and recognize the potential dangers to be watched for and avoided, if it rekindles enthusiasm for the reader's own building experiences and brings them into sharp focus again, the author will feel rewarded and hopeful that his readers will find the same enjoyment and satisfaction in construction purchasing, with its opportunities to meet with hundreds of different personalities, that he has experienced for more than 40 years.

Part I, Chapter 1 describes a proposed building for which the contractor has decided to prepare an estimate and some of the preliminary

[1] National Housing Center Council, *Profile of the Builder and His Industry*, National Association of Home Builders, Washington, D.C.

considerations leading up to the formal submission of a lump-sum proposal. The preparation may involve one or more individuals, with the purchasing function usually applying to subcontractors.

Chapter 2 furnishes details of subcontract bidding and analysis. Chapter 3 involves a study of prospective jobs which offer greater possibilities for awards on a negotiated or cost-plus basis, leading to the preparation of a budget estimate for alteration work. It is further assumed that this contract will require subcontractors to operate on a cost-plus basis and that, because plant operations will have to continue during the construction, several items of work will be performed by the contractor's own employees.

In Part II, Chapters 4 and 5, the negotiations with subcontractors are discussed for both lump-sum and cost-plus work, together with the preparation of agreements for each type of general contract. In Chapter 5, the pricing of items such as ready-mixed concrete, lumber, reinforcing, and masonry materials is discussed for the work which the contractor proposes to perform with his own forces. Situations involving basic material and operating requirements for long-term construction projects are described with some of the differences from the more standardized jobs in metropolitan areas.

Parts III and IV cover a series of topics, including coordination of field operations with purchasing and the relationships of other departments or functions in carrying out the project.

Delivery and expediting of materials and equipment, financial and legal problems that the purchasing man should have knowledge of, policies and procedures to serve as guidelines for handling purchasing or company management in a wider sense, and qualifications and training for those engaged in the work are all discussed.

The final chapter describes the operation of a small department, including the use of departmental forms which establish basic controls for handling subcontract and other procurement work.

John R. Zehner

Acknowledgments

Countless associations over the years in various areas of the country make it difficult to attribute specific ideas to individuals who remain anonymous. Men encountered in construction, engineering, and purchasing positions have always shared their experiences in innumerable ways and helped to create a reservoir of knowledge proving invaluable to anyone handling contractual phases of the building business. To these unnamed individuals I especially express my appreciation for their thoughts which are included in the text.

More recently in its preparation, great assistance has been given by Nelson L. Doe, T. Robert Frost, and Howard A. Clunn, all associates during my work with Turner Construction Company, who have volunteered many helpful suggestions after reading the manuscript draft.

Fred P. Ellison of French, Fink, Markle and McCallion has reviewed and improved the chapter on Legal Problems while G. Harmon Gurney of Madison Square Garden, Inc., has also offered constructive ideas.

Campbell L. Reed, Executive Director of the Building Division of The Associated General Contractors of America; Susan M. Kelsay, The Robert Morris Associates; Frank J. Winters, Executive Vice President of the National Association of Purchasing Management, Inc.; Valerie

Andrews Hale of the American Arbitration Association; Terry L. Peck of The American Institute of Architects; Donald Feiman of Dreier Structural Steel Company; and Patricia Bonnicino of the New York office of the Bureau of Labor Statistics, Department of Labor, have all assisted in the supply and use of information incorporated in the text. John H. Bennett's work in preparing and finalizing the index is much appreciated.

Sincere thanks are extended to William G. Salo, Jr., initially my sponsoring editor at McGraw-Hill, and to Jeremy Robinson, who has carried on with counsel and support.

My thanks also go to Dr. Robert B. Zehner, Chapel Hill, North Carolina, who pondered and puzzled over his father's logic and sentence structure while my wife, Margaret, with great patience, has listened critically to the reading of the manuscript.

Builder's Guide to Contracting

PART ONE

Subcontract Bidding

CHAPTER ONE

Preparation for Bidding:
Requesting and Receiving Bids

For illustrative purposes in this guide, we will assume that the following features apply to the four-story and basement office structure which is to be erected on a 50×250 foot site in Anytown, U.S.A., and for which an owner has requested lump-sum proposals from general contractors operating in the area:

Existing structures
on site .... Two three-story brick and frame houses to be demolished.

Building .... 50×100 feet with balance of site for parking.

Framing .... Structural steel with concrete floors.

Story heights .... Basement to first floor 10 feet; first 12 feet; second to fourth 11 feet.

Exterior .... Brick with granite base course and entrance features; lime-stone sills.

Elevator .... Combination passenger and freight elevator with penthouse on roof.

Toilet rooms .... Ceramic tile floors and walls with plumbing fixtures as shown.

Interior .... Plaster and acoustical work as specified.

Mechanical
and electrical .... Building to be air-conditioned. Electric service to provide for heating, recessed light fixtures, switches, and floor out-lets for electric and telephone facilities.

Plumbing .... Hot and cold water piping, pumps, roof drainage, and toilet fixtures.

On the basis of 25,000 square feet floor area and a median cost of $30 per square foot,[1,*] the building could be budgeted for $750,000. Actually, modest office buildings have been known to cost as little as $15 per square foot in various sections of the country. Recent studies also show costs of $70 per square foot or nearly $1,750,000 for a more-elaborate building in a high-cost area.

Assuming that the $750,000 building cost seems reasonable and is within our contractor's capabilities, that the job appears attractive as a profit-making possibility, and that he stands a reasonable chance in comparison with the other bidders, there are sound grounds for going after the work and letting his subcontractor bidders in the business community realize that he is, in fact, actively interested in submitting a lump-sum proposal and securing this particular project.

On the other hand, should there be too many adverse features about the proposed project which would make it a hazardous undertaking for the contractor, the choice lies between giving it no further consideration or making clear to the owner the reasons for not bidding. Possibly another method of handling could be negotiated. Bad foundation conditions, unusual construction methods, obviously incomplete drawings, or penalty clauses with too short a time for completion can be reasons for passing up an otherwise attractive job.

When the new building plans are nearing completion, the architect usually notifies bidders on general construction that sets of drawings and specifications will be ready for mailing or can be picked up at his office. All bidding documents for a large project should be available for estimating at the same time. From a practical standpoint, large construction organizations are apt to set jobs aside if the indications are that the mechanical and electrical drawings will not be following very shortly. The partial release of information disrupts the estimating and may fail to give sufficient time for bidders on critically important trades. The postponement of the bidding then becomes advisable.

The general contractors are advised when and where proposals should be submitted. In some cases, the bids will be considered privately, i.e., opened by the architect only in the presence of the owner's representative. In such cases, there will not be a public listing of the proposal amounts for the information of others bidding on the job. The contractor or his representative should endeavor to find out from the owner the range and his own position in the bidding. This will be helpful when estimating future work.

Buildings vary greatly due to the nature of their uses and locations, but it is reasonable to expect that 20 or more trades would be required for the type office building described earlier. Additional craftsmen

* Superior numbers refer to Notes at end of chapter.

would be involved for a more-complicated building such as a hospital or laboratory facility. Even the office building, if tenanted by firms requiring special treatment in their spaces, may involve contracting with many subcontractors.

Assembling the estimate may be the work of one or many individuals, depending on the size of the project. As a starting point, it may be fruitful to list the steps in which the purchasing man is particularly involved. If he is also wearing an estimator's hat, other steps will have to be considered more or less simultaneously to meet the bidding deadlines.

ANALYSIS OF SPECIFICATIONS AND BIDDING DOCUMENTS

Our purchasing man should prepare a concise description of the requirements in each trade classification. This summary will be invaluable to anyone trying to visualize the job without resort to rereading the entire specifications. Note where information is given in the specifications *or* drawings but not necessarily in both, even though the responsibility remains to provide a complete installation.

Make note of instances, if any, where the methods of performance and the responsibility for the results are specified. For example, when a contractor does work in accordance with drawings and specifications, for the preparation of which he has not been made responsible, he is generally not chargeable with costs an owner may have to expend to comply with the legal requirements respecting the work or to make the result satisfactory, as, for example, to make a room watertight. It is only when the contractor himself undertakes to produce a designated result that he is bound to produce the result in accordance with the legal and contract requirements.

In some cases, especially where rather technical specifications are involved, the architects may call on sales engineers to outline the requirements. If this is done in a manner, primarily, to restrict other firms from bidding rather than to establish requirements on which other firms may compete, the result may be lessened competition and less opportunity to arrive at a fair price for the product in question.

Note cases where the specified practice is different from the bidder's usual practice, for example, maintaining certain forms for concrete in place for 7 days instead of the usual 3 days. This stipulation would increase lumber requirements and labor costs as well as prolong the job if strictly adhered to. A qualification in the proposal or assurance that the contractor's standard practice would be acceptable should be investigated for this type of clause.

Note where the architect's standard specification has not been corrected for the job in question. In one case, it was stated that the steel detailer (who prepares shop fabrication drawings) was required to visit the architect's office before starting details. The job amounted to 30 tons, and the office was located 700 miles away. It is difficult to estimate how much of such a specification would be enforced.

Sometimes the specifications fail to indicate the trade jurisdictions properly, or they may vary in different parts of the country. In any case, the contractor must see that the cost of each part of the work is covered at least once and not duplicated. The variations in the wage rates of the trades which ultimately perform the work are of less importance than the complete omission of an item.

Occasionally an owner makes an agreement with a subcontractor in advance, with the expectation of having the contractor include the subcontract price in his proposal. Provision for supervising this work should be added and the actual agreement studied to note any additional requirements which the contractor would assume by the assignment.

Where the requirements of a particular trade have not been fully developed, an allowance may be stated in the bidding information to be included, with a contract adjustment to follow for the finalized cost.

Costs for cleaning up after subcontractors should be included if these items are not provided for in the individual trade sections. This applies especially to the specialty subcontractors with large quantities of cartons, crates, and packing cases. It would ordinarily not include cleaning up after subcontractors who have been awarded contracts directly by the owner for miscellaneous work at the conclusion of the job.

A study of the specifications may turn up items on which subcontract bids should be requested in the early stages of estimating. A system of well points for dewatering the site during construction may be desirable and need consideration, for instance, but may not necessarily be specified.

With experience, the estimator will recognize the wide variations in the provisions of the bidding information—some rating, to quote Reuben Samuels of the Thomas Crimmins Contracting Company, as a specification writer's triumph and an estimator's nightmare. For instance, one contract document stated:

Excavation shall include the removal of all materials encountered in excavating, including pavement, concrete foundations, debris, obstructions, sunken materials and all other materials or substances encountered of any kind whatsoever, whether similar or different from those mentioned, whether man-made or natural, whether ordinarily to be expected or not.[2]

The net result of such phrasing serves to raise the bids of all responsible subcontractors but admittedly protects the owner against later claims for extras.

Some specifications lack clarity in failing to indicate exactly those products on which firm prices should be quoted. The statement that a certain product "or equal as approved by the architect" is acceptable permits different interpretations when bidding on a lump-sum basis.

In another instance, it was stated that the work should be done "to the satisfaction of the engineer." Such a personal judgment could vitally affect not only the subcontractor's costs but create a major controversy and result in a delay on the project.

Actually, the engineer, usually engaged by the architect to serve as the structural and mechanical designer of the building for the owner, has a dual responsibility. He must fulfill his technical responsibilities for the design and also act impartially with respect to controversial matters arising in the execution of his portion of the work. The contractor must learn by experience how the engineer—or the architect—interprets these obligations and be guided accordingly.

The estimator in each trade needs to appraise such statements carefully to determine the contingency required to cover similar provisions. For the contractor, the restrictive clauses throughout the specifications need to be recognized and evaluated to determine their relative importance in arriving at the final amount of his proposal.

"Heavy" construction work, such as excavation, or other parts of the building construction industry which are commonly faced with specification clauses allowing the engineer to request the contractor to do one or more things, without compensation, involves a very high degree of risk.

As an example, on a sewer project out for bids, the need for sheeting the trench sides was to be determined by the contractor. One firm judged that there was a 50:50 chance that sheeting would be required. Full sheeting was included on the off-chance that the other bidders would do the same.

When the bids were opened, it became apparent that half of the bidders had figured 100 percent sheeting and that half had figured none. Two weeks after the project was awarded to the low bidder, there was a cave-in resulting in death and injury. The town engineer, state labor department, and the contractor's insurance company immediately demanded that the remainder of the trench be tight-sheeted. The contractor complied and went bankrupt.[3]

All this could have been avoided if the contract documents had made an equitable provision for a contingency of this nature. In the case mentioned, had the contractor included 50 percent of the sheeting cost, the bid would have been noncompetitive and borne out the conclusion of the staff that this particular job should not be bid at all.

MISSING INFORMATION

Note apparent errors in the drawings, specifications, or schedules. In the case of one housing development, the reinforcing schedule for a type of slab erroneously stated "vermiculite-conc." A federal district court in Delaware held that this was an obvious drafting error and that the contractor was liable for furnishing the reinforcing required in the slabs. He should have exercised prudence and requested an interpretation or included an allowance for proper reinforcing in his bid.[4]

In another instance, the finish schedule noted below caused confusion and protracted litigation due to different interpretations of the requirements:

Finish number	Concrete walls and columns	Hollow tile or brick walls
1	Cement plaster	Cement plaster
2	Gypsum plaster	Gypsum plaster
3	Cement plaster	Cement plaster
4	Concrete plywood forms	Cement plaster
5	Concrete plywood forms	ditto

The plastering subcontractor testified that he had prepared his bid assuming that "ditto" applied to the concrete plywood forms; i.e., he had read item #5 horizontally. After several years, a New York appellate court held for the contractor—in that "ditto" applied to the plastering finish #5. Interestingly, the court's 3 to 2 decision found two judges considering that the schedule should have been read horizontally.[5] It was another case where more thought when estimating in regard to the actual appearance might have sensibly suggested including an allowance for plaster, thereby avoiding expensive litigation.

Not infrequently, architectural drawings do not show full information needed by mechanical and electrical bidders. Sets of architectural drawings should be made available to these firms if necessary and instructions given if there are discrepancies. Similarly, other trades, such as concrete and miscellaneous iron bidders, will need architectural and structural drawings. It is highly desirable that *all* drawings and specifications be considered as part of the bidding information, even though a specific bidder may not be concerned with all the data. In all cases, subcontract bidders should be encouraged to submit their questions to the contractor to be resolved by the architect at the earliest date so that corrected information may be issued in bulletin form to all concerned.

While there is no longer a general professional approval of specifications stating that the contractor is to furnish all items necessary to complete a structure or any part of it, regardless of whether or not it is called for on the drawings or specifications, the fact remains that some bidding documents still include such clauses. The prudent contractor estimating a job should endeavor to obtain some ground rules on how this would be interpreted in actual practice.

PREPARING THE BIDDING LIST

In order to receive three bids for a particular trade, the contractor may need to invite five or six subcontractors in the vicinity to bid. It must be recognized that their work schedules may not permit spending estimating time when bids are being requested. Those who are invited to bid should be qualified, based on the contractor's past experience or prior checking.

General contractors normally compile a list of subcontractors in each trade, noting their addresses, telephone numbers, and special qualifications in a loose-leaf book or circular recording file. Each entry is categorized so that by glancing through the names selections can be made for the particular job. Subcontractors usually forward a résumé of their qualifications when desiring to be placed on a general contractor's bidding list. This will help in checking experience and rechecking selections.

CONSIDERATIONS IN SELECTING BIDDERS

There may be times when an inquiry to subcontractors is necessary to determine prospective bidders. Information on experience, work load, and references will help. A typical letter might request:

1. A list of buildings on which work has been completed during the past 5 years, with approximate subcontract amounts together with references who may be contacted among contractors, architects, owners, and material suppliers.

2. A list of current work with its status and completion dates as well as future subcontracts to be performed.

3. A description of the subcontractor's organization and the personnel who would be available to handle work and their qualifications. Is there a responsible assistant in the organization, or does the subcontractor operate pretty much as a one-man show? What would happen if he became ill or incapacitated?

4. Trade union affiliations.

5. Banking references and/or a recently audited financial statement.

The contractor will recognize that subcontractors "grow" and broaden their operations over the years. At the same time, depending on the type of building involved, one would ordinarily expect that contractors whose experience has been with more finished types of buildings might be better qualified than those whose work has been limited to strictly speculative structures with a somewhat dubious class of workmanship.

Finally, does the subcontractor appear to be a specialist who is thoroughly experienced in his own business, or is there a tendency to spread his efforts into other activities to the detriment of the principal occupation?

After a few contracts have been completed and rating cards have been prepared for the subcontractors (Fig. 21), it becomes possible to make a more-accurate determination based on actual experience. Scanning the cards will facilitate a separation of the most acceptable bidders as well as those to be avoided when soliciting bids.

Many contractors perform their own carpentry, concrete, or masonry work. Obviously if manpower is available with proper supervision and capital, there is no compelling reason to award subcontracts for these trades unless low bids have been received for them after the work has been estimated in detail by the contractor's staff. Normally bids would not be requested on those trades where the contractor has no intention to award subcontracts. Subcontractors expect that if bids are requested, the contractor will eventually make an award to one of the bidders.

The subcontractor who has idle equipment would probably prefer to operate rather than have it stand idle. This could result in wide ranges of pricing in the trade. Where this situation exists and very low prices are received, particularly if rock excavation is involved, the contractor might well decide to subcontract the work and avoid taking on the responsibility for the specialized blasting required.

It is well to broaden the list of bidders occasionally and not use the same ones repeatedly if more estimating work is expected in the near future.

Qualified subcontractors who come to the area performing work for national organizations, and who are able to handle additional local work advantageously, should not be overlooked. This applies particularly to painting, floor covering, and other architectural and finishing trades. Where there is a dearth of responsible bidders a concerted effort will have to be made to avoid too much dependence on what may become a tight monopoly control in the field.

RED FLAGS TO WATCH FOR

All contractors may not be equally selective in requesting subcontract bids. Consequently, there may be incompetent or unreliable bidders broadcasting erroneous figures which can cause confusion and possible later litigation involving other contractors who are bidding. Using such estimates tends to establish incorrect bases for judging the amounts submitted by other subcontractors.

If the bidding is complex, consideration should be given to requesting subcontractors to submit their proposals on a form which is sent them and follow the practice of the general contractor in his submission, as is sometimes done when proposal forms are included in the bidding information; or bidders should be requested to follow the draft in preparing their estimates on their own letterheads. Bid forms simplify the analysis of estimates and make it possible to spot irregularities more easily. This applies especially where unit prices and alternates are involved.

On cost-plus work it is customary to request duplicate or even triplicate copies of bids if owners and architects are involved in their original consideration. It might be noted that the experienced owner will generally see that all bids remain in the custody of the contractor if he is to remain responsible for the actual negotiation of the subcontract.

The contractor may recognize that local-area subcontractors may not be as effective as others he has known in other parts of the country. Office procedures, especially in the mechanical and electrical trades, may be improved by a joint venture, with a competent subcontractor taking the lead in office coordinating and the local subcontractor procuring the labor supply. The experienced purchasing man can act as the catalyst in bringing these firms together for bidding purposes.

Most contractors are inclined to give more credence to the listing of subcontractors in trade registers than to the listings in the yellow pages of telephone books, which are apt to include less-experienced operators who may or may not have sufficient training in the building field. Telephone listings do not differentiate between union and non-union bidders. Some competent subcontractors specialize in working directly for owners but lack experience in working with other subcontractors and a general contractor on new-building construction.

In selecting bidders if the job will be operated on a union basis, the purchasing man must be careful not to invite nonunion bidders to submit proposals. If the job will be operated on an open-shop basis, with both union and nonunion subcontractors working on the same project, it becomes extremely important to have the ground rules established and agreed upon before any subcontracts are awarded.

Union and nonunion subcontractors knowingly will not try to bid for the same job. If bids are invited from both, negotiations may be carried on for a while and even lead to subcontract awards which could result in labor picketing or delays in prosecuting the work until the situation is resolved.

TABLE A-3 Construction-Worker Employment in Contract Construction by Industry and Occupation (000) for 1969

	General building contractors	Highway and street	Other non-building	Plumbing and heating	Painting and decorating	Electrical work	Masonry, stone and tile	Roofing and sheet metal work	Other special trades	All contract construction
Carpenters	308.1	16.7	19.4	24.1	3.0	8.7	19.2	14.6	106.3	520.5
Brickmasons and other stone workers	48.1	1.2	1.6	0.0	0.0	0.0	96.6	0.0	8.0	155.8
Cement and concrete finishers	12.4	6.0	8.0	0.0	0.0	0.0	0.0	0.0	32.2	58.7
Electricians	11.9	0.0	0.0	2.4	0.0	142.6	0.0	0.0	0.0	157.0
Excavating and road machine operators	19.7	75.6	54.5	0.0	0.0	0.0	0.0	0.0	24.4	174.5
Painters and paperhangers	64.3	1.5	2.0	0.0	94.6	0.0	0.0	0.0	0.0	162.5
Plasterers	0.0	0.0	0.0	0.0	0.0	0.0	18.0	0.0	0.0	18.0
Plumbers and pipefitters	18.2	0.0	4.7	124.0	0.0	0.0	0.0	0.0	0.0	146.9
Roofers and slaters	0.0	0.0	0.0	0.0	0.0	0.0	0.0	51.3	0.0	51.3
Structural metal workers	0.0	2.6	13.9	0.0	0.0	0.0	0.0	0.0	21.7	38.3
Craftsmen, foremen	113.8	28.2	56.2	46.5	4.7	21.0	9.2	4.6	65.3	349.9
Truck drivers	40.3	30.7	70.5	8.5	0.0	10.3	0.0	0.0	10.7	171.2
Operatives, service workers	45.3	22.0	27.8	36.9	3.2	15.3	10.1	5.1	42.0	208.0
Laborers	178.8	97.0	93.6	76.8	7.2	27.7	55.1	21.3	107.7	665.8
Total	861.3	281.9	352.8	319.5	112.9	225.9	208.5	97.1	418.8	2879.0

TABLE A-4 Construction Worker Union Employees by Industry and Occupation (000) for 1969

	General building	Highway and street	Other non-building	Plumbing and heating	Painting	Electrical work	Stone and tile	Sheet metal work	Other special trades	All workers	Union to total*
Carpenters	141.9	10.0	15.1	20.5	1.6	8.2	15.5	10.0	100.0	322.8	0.620
Brickmasons and other stone workers	30.5	1.0	1.5	0.0	0.0	0.0	77.5	0.0	7.2	117.8	0.756
Cement and concrete finishers	5.8	3.4	5.8	0.0	0.0	0.0	0.0	0.0	27.4	42.5	0.724
Electricians	8.3	0.0	0.0	2.1	0.0	135.8	0.0	0.0	0.0	146.2	0.931
Excavating and road machine operators	15.0	53.0	45.0	0.0	0.0	0.0	0.0	0.0	22.5	135.5	0.777
Painters and paperhangers	34.1	1.2	1.8	0.0	68.0	0.0	0.0	0.0	0.0	105.2	0.647
Plasterers	0.0	0.0	0.0	0.0	0.0	0.0	14.4	0.0	0.0	14.4	0.800
Plumbers and pipefitters	10.3	0.0	3.9	103.4	0.0	0.0	0.0	0.0	0.0	117.5	0.800
Roofers and slaters	0.0	0.0	0.0	0.0	0.0	0.0	0.0	27.5	0.0	27.5	0.535
Structural metal workers	0.0	1.6	10.5	0.0	0.0	0.0	0.0	0.0	18.5	30.5	0.797
Craftsmen, foremen	57.8	18.7	46.4	41.0	2.0	19.6	6.0	3.1	53.6	248.2	0.709
Truck drivers	17.3	10.4	47.9	4.5	0.0	6.0	0.0	0.0	5.1	91.2	0.532
Operatives, service workers	19.6	11.5	22.1	29.7	1.3	13.5	6.0	3.0	30.3	136.9	0.658
Laborers	60.4	35.9	59.0	43.8	2.0	16.9	23.2	9.7	55.1	306.0	0.406
Total	401.2	146.8	258.9	245.0	74.9	199.9	142.7	53.3	319.7	1842.3	0.640
Ratio Union/Total*	0.466	0.521	0.734	0.767	0.663	0.885	0.685	0.549	0.763	0.640	

* This is the proportion of union employees in the construction work force.

13

Union control over trades in the construction industry is so well established, especially in many metropolitan areas, that there is a tendency to accept it as one of the facts of life (Tables A-3 to A-9).[6] Actually, in 1969, 64 percent of the workers employed in contract construction were union members and 36 percent were nonunion (Table A-4).

Other charts giving an insight into the situation with respect to union and nonunion construction follow:

TABLE A-5   Membership of Unions Affiliated with the Building and Construction Trades Department, 1968

Unions	Total membership 1968	Reported membership outside U.S.	Percent in construction
Asbestos workers.................	16,698	1,812	†
Boilermakers....................	140,000	9,030	20
Bricklayers.....................	160,000	5,355*	†
Carpenters.....................	793,000	76,962	70
Electrical (IBEW)..............	897,114	54,389	19
Elevator constructors...........	15,633	2,226	100
Operating engineers............	350,000	21,570	70
Granite cutters.................	3,300		
Iron workers....................	167,928	12,308	74.5
Laborers.......................	553,102	36,000	80
Lathers........................	16,007	1,454*	†
Marble setters..................	8,206	300	84
Painters........................	200,000	10,006	†
Plasterers and cement masons.....	68,000	3,000	99
Plumbers.......................	297,023	30,416	†
Roofers........................	24,729	......	100
Sheet metal workers.............	140,000	13,692	†

* Data obtained from "Labour Organizations in Canada," 1968 ed., Dept. of Labour, Ottawa, Canada.

† Data not reported.

Source: U.S. Department of Labor, Bureau of Labor Statistics, February, 1970.

Those comparing more comprehensively the statistics for union and nonunion employment in the construction industry will find a wide variation in figures which may be due to the methods of assembling the data and the different concepts, classifications, and coverage in making the presentation.

Report 417, entitled "Selected Earnings and Demographic Characteristics of Union Members, 1970," prepared by the U.S. Department of Labor through its Bureau of Labor Statistics and published in October 1972, notes that in some cases the labor union reports include in their totals retirees, unemployed members, and individuals in the Armed Forces.

TABLE A-6 Average Hourly Earnings ($) for Union Construction Workers (1960–1969)

	1960	1961	1962	1963	1964	1965	1966	1967	1968	1969
Carpenters	3.71	3.85	4.00	4.14	4.29	4.48	4.68	4.94	5.33	5.77
Brickmasons and other stone workers	4.12	4.28	4.42	4.54	4.67	4.79	4.99	5.21	5.61	6.04
Cement and concrete finishers	3.63	3.79	3.91	3.43	4.15	4.34	4.53	4.79	5.09	5.51
Electricians	3.98	4.14	4.33	4.50	4.64	4.78	4.95	5.18	5.52	6.06
Excavating and road machine operators	4.20	4.37	4.53	4.70	4.87	5.06	5.27	5.57	5.94	6.46
Painters and paperhangers	3.51	3.64	3.81	3.95	4.09	4.26	4.44	4.67	4.96	5.36
Plasterers	4.03	4.11	4.24	4.36	4.55	4.64	4.82	5.03	5.30	5.69
Plumbers and pipefitters	4.01	4.16	4.32	4.49	4.66	4.85	5.04	5.32	5.69	6.19
Roofers and slaters	3.69	3.85	4.01	4.16	4.31	4.48	4.69	4.96	5.34	5.78
Structural metal workers	3.89	4.06	4.24	4.39	4.53	4.69	4.93	5.20	5.56	6.01
Craftsmen, foremen	3.90	4.05	4.21	4.36	4.51	4.69	4.89	5.16	5.51	5.99
Truck drivers	2.71	2.81	2.92	3.05	3.17	3.28	3.39	3.59	3.78	4.06
Operatives, service workers	3.42	3.56	3.70	3.83	3.97	4.13	4.30	4.54	4.84	5.27
Laborers	2.72	2.84	2.95	3.06	3.18	3.33	3.49	3.67	3.92	4.20
Total	3.57	3.72	3.86	4.00	4.14	4.30	4.51	4.76	5.09	5.50

TABLE A-7   Average Hourly Earnings ($) of Nonunion Construction Workers (1960–1969)

	1960	1961	1962	1963	1964	1965	1966	1967	1968	1969
Carpenters....................	2.60	2.70	2.80	2.90	3.00	3.13	3.27	3.46	3.73	4.04
Brickmasons and other stone workers.	2.94	3.06	3.16	3.25	3.33	3.43	3.57	3.72	4.01	4.32
Cement and concrete finishers......	2.33	2.43	2.51	2.20	2.66	2.78	2.90	3.07	3.26	3.53
Electricians..................	2.52	2.62	2.74	2.85	2.94	3.03	3.13	3.28	3.50	3.84
Excavating and road machine operators.................	2.80	2.91	3.02	3.13	3.24	3.37	3.52	3.71	3.96	4.31
Painters and paperhangers........	2.26	2.34	2.45	2.54	2.63	2.74	2.85	3.01	3.19	3.45
Plasterers....................	2.80	2.85	2.94	3.02	3.15	3.22	3.34	3.47	3.67	3.94
Plumbers and pipefitters.........	2.96	3.07	3.19	3.31	3.44	3.58	3.72	3.92	4.20	4.57
Roofers and slaters.............	2.42	2.52	2.63	2.73	2.83	2.94	3.08	3.25	3.51	3.79
Structural metal workers.........	3.10	3.23	3.38	3.50	3.61	3.73	3.93	4.14	4.42	4.78
Craftsmen, foremen.............	2.52	2.62	2.72	2.82	2.92	3.04	3.17	3.34	3.65	3.88
Truck drivers.................	1.98	2.05	2.13	2.23	2.31	2.39	2.48	2.62	2.76	2.97
Operatives, service workers.......	2.43	2.53	2.63	2.72	2.82	2.93	3.05	3.22	3.44	3.74
Laborers.....................	1.68	1.76	1.82	1.89	1.96	2.06	2.16	2.27	2.42	2.59
Total...................	2.19	2.29	2.38	2.47	2.55	2.66	2.81	2.96	3.16	3.40

TABLE A-8 Contract Construction
Average Hourly Earnings Estimated ($) of Union Construction Workers by Type of Construction (1960–1969)

	General building contractors	Highway and street	Other non-building	Plumbing and heating	Painting and decorating	Electrical work	Masonry, stone and tile	Roofing and sheet metal work	Other special trades	All contract construction
1960	3.56	3.51	3.36	3.61	3.50	3.77	3.76	3.47	3.55	3.57
1961	3.71	3.67	3.51	3.76	3.63	3.95	3.91	3.63	3.70	3.72
1962	3.84	3.80	3.63	3.89	3.80	4.13	4.04	3.82	3.84	3.86
1963	3.98	3.98	3.78	4.05	3.94	4.30	4.15	3.92	3.94	4.00
1964	4.11	4.10	3.90	4.19	4.08	4.39	4.30	4.08	4.11	4.14
1965	4.29	4.27	4.07	4.35	4.25	4.53	4.46	4.23	4.29	4.30
1966	4.49	4.50	4.28	4.59	4.43	4.74	4.64	4.43	4.51	4.51
1967	4.74	4.74	4.51	4.85	4.67	4.99	4.88	4.71	4.75	4.76
1968	5.08	5.07	4.81	5.16	4.96	5.31	5.23	5.08	5.09	5.09
1969	5.49	5.50	5.22	5.61	5.35	5.77	5.64	5.47	5.51	5.50

TABLE A-9 Contract Construction
Average Hourly Earnings Estimated ($) of Nonunion Construction Workers by Type of Construction (1960–1969)

	General building contractors	Highway and street	Other non-building	Plumbing and heating	Painting and decorating	Electrical work	Masonry, stone and tile	Roofing and sheet metal work	Other special trades	All contract construction
1960	2.29	2.06	2.09	2.19	2.20	2.05	2.20	2.20	2.02	2.19
1961	2.39	2.16	2.18	2.28	2.29	2.17	2.31	2.31	2.12	2.29
1962	2.48	2.24	2.26	2.36	2.39	2.27	2.38	2.46	2.20	2.38
1963	2.57	2.34	2.36	2.48	2.49	2.37	2.45	2.50	2.28	2.47
1964	2.65	2.42	2.45	2.55	2.58	2.41	2.55	2.61	2.37	2.55
1965	2.77	2.52	2.55	2.65	2.69	2.50	2.67	2.70	2.48	2.66
1966	2.91	2.67	2.70	2.84	2.82	2.65	2.79	2.83	2.63	2.81
1967	3.07	2.80	2.83	3.00	2.96	2.79	2.95	3.02	2.75	2.96
1968	3.29	3.00	3.02	3.17	3.15	2.96	3.14	3.27	2.94	3.16
1969	3.55	3.23	3.26	3.44	3.40	3.17	3.39	3.50	3.16	3.40

Table 7 of the above report notes that 38.3 percent of those in the construction industry are considered union members with median earnings (1970) of $11,213, while the comparable figure for those not in labor unions was $7,282.

The report should be studied in its entirety for further data on occupations, sex, race, and region. The analysis, data, and recommendations given are especially recommended for all to whom the union-nonunion relationship is a matter of concern.

Other tables published periodically during recent years in *Engineering News-Record* indicate labor rates for various cities in the country.   These tables do not show the proportion of union workers in comparison with the total work force or the differentials between union and open-shop operations.   The higher rates generally apply in Northern cities where union operations are more prevalent.

Significantly, union rates in many of the smaller locals outside the major population centers tend to be lower.   The wage stabilization efforts of the Construction Industry Stabilization Committee have helped to curtail the runaway wage increases which prevailed some years ago.

ISOLATED JOBS AND NEW RELATIONSHIPS

Securing sufficient competent bidders is a continuing task not only in the metropolitan areas but even more so where the projects are isolated.   On the one hand, it points up the need for fair treatment of subcontractors so that they will be interested in bidding and handling more of the contractor's work.   On the other hand, by developing a knowledge of pricing and the requirements for the various trades from the subcontractors, the contractor maintains his competitiveness as well.

Better coordination of jobs by a new contractor coming into an area will gradually overcome the occasional preference by some subcontractors to deal exclusively with contractors they have known.   Giving all subcontractors a square deal will always pay good returns.

ORDER FOR DRAWINGS AND SPECIFICATIONS

Most specifications state that two sets of bidding information will be loaned without charge to general contractors submitting proposals. Additional sets are normally available for purchase.   Usually one set will be used exclusively by the estimator, and the other will be handled by the purchasing man for his analysis and for study by the various subcontractors in the contractor's plan room.

To secure good bidding coverage, the contractor will usually have to buy some additional sets at considerable cost if many drawings are involved. Some key trades may require the drawings for several days, while others will be able to use those in the contractor's plan room, in other contractors' offices, or in other cities where the architect has provided bidding information. Naturally, the aim is to use the sets as efficiently as possible. Although risky even under close control, sets can be taken apart and applicable sheets loaned to several subcontractors in different trades. Extra portions of the specifications can be secured through photocopying.

It is well to double-check information which other general contractors have furnished to subcontractors not on the contractor's original bidding list. One kitchen-equipment subcontractor found (to his regret), several months after the award for a large veterans' hospital in Kansas, that the bidding data furnished by an unsuccessful bidder on general construction—which he had assumed to be complete—did not include drawings for all the buildings involved. It resulted in a very substantial loss for which the court allowed no relief.[7]

CONTRACTOR'S SUPPLEMENTARY SPECIFICATIONS FOR ALL SUBCONTRACTORS

When preparing to request bids for either lump-sum or cost-plus general construction, it is customary for the contractor to prepare additional specifications (Appendix A) to supplement the detailed or sometimes outline bidding data which the architect has furnished. The information would primarily include the operational guidelines to be followed on the job with which all subcontractors would be expected to comply. Sometimes these data are included in the invitation to bid. More often it will be a separate document, dated and referred to in the bidding information.

As subcontracts are awarded, additions or corrections may be made to the supplementary specifications, carrying a revision date applicable to the subcontract in question.

In preparing these additional provisions, which should carry a title to distinguish them from the architect's bulletins, the contractor will avoid inclusion of superfluous items for the particular project.

INVITATIONS TO BID

Post card invitation to bid notices (Fig. 1) to five or six bidders for each of the 20 or more trades should be mailed as promptly as possible to the selected subcontractors for each trade. This information may be supplemented with a letter or card if required.

After subcontractors have been invited to bid, there should be no objection to revealing their names.   The bidders could obtain the information easily by contacting material suppliers.   Should they decide not to bid because the competition is not to their liking and drop out, there should still be time to secure a replacement in the particular trade involved.

```
┌─────────────────────────────────────────────────────────────┐
│ A.B.C. Construction Company              (216)697-1000        │
│ Brown Street, Cleveland, Ohio 44101       July 22, 1974       │
├──────────────────────────────────┬──────────────────────────┤
│ You are invited to bid on        │ Office Building          │
│ work in your line for:           │ Fremont, Ohio            │
├──────────────────────────────────┴──────────────────────────┤
│ General Contractors' bids due Thursday, August 15, 1974.     │
│ Please submit your bid by                                    │
│                   ┌──────────────┐                           │
│                   │ Wednesday,   │                           │
│                   │ August 14    │                           │
│                   └──────────────┘                           │
├─────────────────────────────────────────────────────────────┤
│ Note:  Please submit alternate and unit prices.  Note        │
│ broad form hold harmless clause in subcontract form          │
│ available at our office for examination.  Kindly             │
│ acknowledge that you will bid.                               │
│                                              J. J. Smith      │
└─────────────────────────────────────────────────────────────┘
```

FIG. 1   Invitation to Bid (post card form).

Assuming either that some building materials may be required or that the contractor would like to have current prices available for checking purposes, the estimator would normally furnish a list of quantities which the purchasing man will forward to various vendors.   These prices may be needed in some cases for quick estimates if subcontractors fail to submit bids at the last minute.

BID VERIFICATION

The request for acknowledgment appearing on the post card will usually bring an indication of whether the bidder intends to submit an estimate. Sometimes the subcontractors are busy at the time, and nothing happens for a few days.   The contractor would be well advised to verify what the subcontractor's intentions are: if he has already found that more information is needed, if he is not bidding, or if a decision on bidding will be made shortly.   This will help in routing drawings which may be loaned for bidding purposes.

At this stage, contractor's bulletins based on the architect's interpretations should be issued to clear up questions for all bidders.   The sooner the bidders study the drawings and start asking questions, the sooner the various uncertainties will be resolved.   Informational bulletins issued through purchasing should have a distinctive designation so that they will not be confused with architect's bulletins, similarly numbered.

BID TABULATION FORMS

As bids are received, they are attached to the appropriate trade forms and major features listed on the bid tabulation form, or quotation sheet (Fig. 2).   Prices, timing, exceptions or exclusions, and unit prices should be noted.   Every effort should be made to set up tabulations so that the bid amounts indicated are comparable.   The forms should be scanned frequently to see where bids are missing and where telephone calls will be necessary to produce results in getting a complete and useful comparison of all bids.   Even when incomplete bids are offered, it may be possible to make worthwhile comparisons with portions of other bids being submitted.

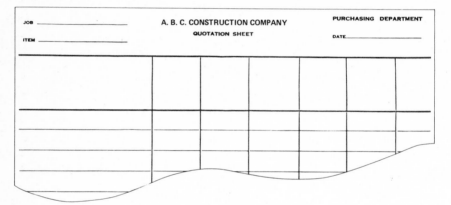

FIG. 2   Bid Tabulation Form, or Quotation Sheet.

It should be noted that some subcontractors will find it possible to take off quantities and submit their bids well in advance of the due date.   If new information is issued during the interim, the bidders will have to be asked to check the new data and possibly submit a new bid.   Care must be taken to have all bids consistent on the basis of final requirements.

If the jobs are larger, with many alternates, unit prices, and the possibility of partial estimates, it may be advantageous to prepare a special tabulation sheet for the job so that the status of the various bids can be determined more readily.   This same sheet can be used as the estimate is being prepared to reflect the reasoning in establishing the estimate for the trade and any discount or contingency.

CHECK SHEETS

Usually there are last-minute items of information or corrections phoned in by subcontractors during the preparation of the estimate.   It is good policy to have check sheets, or telephone bid sheets (Fig. 3), mimeographed for the particular job, with spaces to confirm the bidding data, prices, bulletin numbers and addenda, alternates, unit prices, and so forth, as well as exceptions, if any. Information should be noted regarding inclusion of sales or use taxes and insurance premiums in exact accordance with the specified limits of coverage. On jobs having tax-exempt status, it is well to verify that these taxes are not included. Similarly, on jobs where the owner would carry all insurance costs under

A. B. Jones Company		Aug. 1, 1974
Fremont, Ohio		Telephone Bid Sheet

WORK INCLUDES:

_____ Trade   Spec. Sect. #

_____

_____ Company   Phone No.

_____

_____ Person on Phone   Taken By

_____ BASE BID $ _____

_____ _____ Complete per Plans & Specs

_____ _____ A.B.C. Advisement No. 1

WORK EXCLUDES: _____ Per Recording Sheet Checklist

_____ _____ Includes Sales Tax

_____ _____ Includes Erection

_____ _____ Includes Hoist Facilities

_____ _____ Includes Temporary Requirements

_____ UNIT PRICES:

Numbers	Unit	Add	Deduct

Alternates:
_____

_____

FIG. 3   Check Sheet, or Telephone Bid Sheet.

a "wrap-up" policy, the subcontractor's insurance costs should be excluded.

A partial receipt of information results in loss of time when minutes are essential. Advance preparation of forms makes it possible for less-experienced workers to ask for and receive a complete story. Confirmation of telephone bids should always be requested. The job may "come to life" a few months later and written bids become very helpful—quite often necessary to avoid misunderstanding.

ESTIMATE COORDINATION

While the handling of subcontract pricing has been proceeding, the chief estimator, or person so assigned, has been fully occupied with coordination of various phases of the work to submit a proposal in proper form. Others, perhaps the superintendent to be assigned to the job, will have made a study of the temporary construction require-ments—the temporary light and power, heat, water, and sanitary facili-ties. Hoist towers, sidewalk bridges, walkways, fencing, truck access through the building with possibly heavier steel, and stronger concrete slabs in lieu of protective planking may require pricing through subcon-tractors who are bidding.

Once these decisions have been made, all bidders should be informed by bulletin, noting the temporary facilities which would be provided to them under a subcontract, including also the rates to be charged for any hoisting facilities or other items furnished for their use. The total of these hoisting charges will reflect a credit against the contractor's gross installation, rental, and operating costs of the hoisting operation. These installations, with their operating engineers, constitute an expen-sive part of the total plant costs.

It sometimes happens that by overtime use of the hoist there will be sufficient operating hours for the hoist to accommodate the needs of all the subcontractor's trades on the job, or at least to minimize the total number of hoist installations.

Many subcontractors handling masonry or plastering materials find that by operating on overtime they can achieve a greater efficiency in their costs of stocking floors. By establishing one rate for either straight or overtime usage, the contractor will usually find that the subcontractors have no objection to paying the overtime for their own employees, and the need for additional hoisting facilities may be dispensed with.

Perhaps there is special equipment which could be used on the job for which proposals should be obtained and applicable rental charges determined for inclusion in the estimate.

SITE SURVEY

Another critical feature is the preparation of the site survey to determine contingencies to be provided for. In addition to a study of temporary construction needs, this survey will develop data of interest to subcontractors; as soon as information is available, it should be transmitted to the bidders.

Plans of buildings to be demolished, construction drawings of nearby buildings, and rock excavation depths in the vicinity are all typical items which may be determined during the survey.

While the subcontractors are concerned with the quantities of materials to be handled, the manner in which the work will be performed, and other conditions revealed by the survey, it is the contractor who ultimately will have to accept responsibility for the complete project and the shortcomings, if any, of his subcontractor bidders.

Checklists prepared by architects' and contractors' associations are available and can be valuable both for specifications and for data filing and cost accounting.[8] It should be noted, however, that while job specifications may be specific, there are variations in trade practices throughout the country so that all assumptions need verification. For instance, utility companies in some cases will include and install overhead services or even underground services to the building, while in other cases their service stops at the property line.

SCHEDULING

A schedule for the performance of the work will have to be prepared giving starting and completion dates for the trades, which will be used in support of the study of back-page or overhead charges referred to in Chapter 2. Major subcontractors bidding on the pace-setting trades such as rock excavation, foundations, and structural steel can give a good clue regarding expected progress.

Should there be trades that might significantly affect progress because of delivery problems (such as imported stone, special ornamental metal, or air-conditioning equipment), the necessary data should be obtained from the bidders or vendors by the purchasing man and considered in setting up the commencement and duration of their work.

Scheduling of a job starts with the selection of a target date, generally through comparison with performance on similar buildings. With experience and recognition of past problems and the practical means to overcome them, the traditional bar chart method not only is completely feasible but also is usually accurate enough for estimating purposes. (See Fig. 4.)

Where the projects are larger operations with new trades and different conditions, critical path method scheduling (CPM) is fast becoming a necessary adjunct for coordinating performance, evaluating costs, and effecting completion.

If the job is in the several million dollar range, many owners will request that these CPM schedules be prepared after an award, in which case the cost of preparation should be included in the estimate.

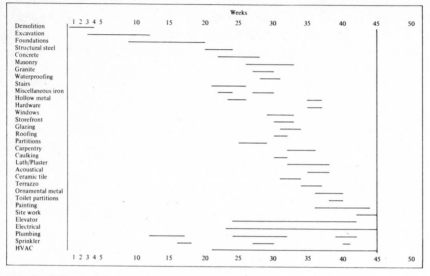

FIG. 4   Estimated Construction Schedule.

There will undoubtedly be increased use of CPM scheduling, even on smaller jobs. For the purchasing man looking toward the future, a familiarity with this subject, and its relationship to subcontractors and the data which they are expected to furnish, is most worthwhile. Only a comparatively small percentage of contractors presently use CPM scheduling. Recent figures for those who do, indicate the following:[9]

Volume of business	CPM contractors (%)
Over $10,000,000 per year	22
Over $1,000,000 per year	56
Nearly $1,000,000 per year	11
Less than $500,000 per year	11
	100

OUTSIDE ESTIMATING SERVICES

As the contractor considers the pros and cons of submitting a proposal, one factor studied will be the availability of his own staff at a time when pressing problems on jobs underway may demand his attention.

In some areas an estimating organization or retired estimator can furnish figures needed. Frequently subcontractors for trades such as miscellaneous iron, reinforcing, concrete, formwork, and masonry will secure quantities from the estimating service and use their own cost records for pricing the work.

In the long run, the contractor will know his job better if the quantities are established by his own people and the outside service is employed only in an emergency.

Square foot quantity estimates, whenever prepared, will be useful in comparing and checking items included in bids of subcontractors for roofing, acoustic ceilings, resilient floors, plaster- or cement-finished ceilings, painting, or various types of wall-construction materials or finishes.

MANAGEMENT'S RESPONSIBILITIES

The contractor will normally investigate the availability of the labor supply personally. However, it is a matter of concern to those purchasing, since the scheduled completion will be in jeopardy and costs will increase if men are not available. Current wage rates, established increases, and fringe benefits will be noted and projected through the period when the work is to be performed.

There are alternatives for labor shortages, but one cannot be assured of a satisfactory solution. For instance, on a recent New York City job involving a large quantity of special roofing tile, it was necessary to handle the job as a joint venture because neither of the two larger roofing subcontractors had access to sufficient slaters to perform the work within a reasonable time.

On occasion, the business agent of a union will arrange for or agree to have union mechanics report on a permit basis from other areas of the country; or, much less frequently, he will admit his inability to furnish sufficient men and allow union men of other trades to assist in the work. Such conditions may also result in expensive overtime costs and, possibly, travel allowances.

In some states, with individual craft unions on strike, there is a tendency for members to work at other trades, resulting often in a prolongation of the strike.

The high rates of union wages and union trade practices are creating situations in several areas of the country where nonunion, or open-shop, work is becoming more prevalent. It cannot be assumed, however, that a job which has started on a craft-union basis can be completed on a less-costly nonunion basis, even though the nonunion labor supply has become sufficient. In any case, the shortage of men delays progress on jobs, and the vicious circle of increasing costs continues because a greater proportion of work may require completion in a period with escalated wages.

The period of the year may have considerable effect on the ability of the subcontractors to round up sufficient men. For instance, school-building construction and fall openings may deplete the force of available men during the summer months, making overtime a necessity since school construction normally involves tight scheduling.

Advertising programs firmed up well in advance of the grand opening of a store can require expensive overtime if there is not sufficient slack in the schedule, to say nothing of storage and rehandling of merchandise that must be shipped to be ready for opening day. Easter, Thanksgiving, and Christmas are all periods when sales promotions affect construction programs that are underway. At times the store-fixture installations, perhaps by the owner's subcontractors, will take priority and interfere with the progress of the contractor's people. Stronger subcontractors will usually be able to ride out the crash program, while the inexperienced will have difficulty in keeping to schedule. Sports facilities similarly have target dates which must be maintained.

It is a period when changes are frequently made. All subcontractors need to be conscious of the need to give estimates promptly and secure authorization to proceed from the contractor. In turn, the contractor should satisfy himself that the owner fully understands the additional commitments being undertaken and is financially able to carry through.

This points up the need to try to determine when a particular trade is needed and whether, *at that time,* and considering other projects as well as possible delays, there will be a sufficient supply of labor. Information given to bidders about starting and completion dates is therefore extremely critical. Care must be taken not to promise a certain rate of progress by subcontractors in a manner that causes the contractor to be held responsible for the delays of his subcontractors and additional expenses caused by others.

On the other hand, the contractor and estimator will recognize that a prolonged delay in the award of the general contract creates a situation which needs to be explored with all major subcontractors concerning wage escalation they may be faced with because of the delays. Under such circumstances, an agreement should be made with the owner for compensation, prior to contracting.

FINANCING

Most contractors, as well as subcontractors, will not have sufficient working capital to finance any but a small project without relying on advances from some lending agency. The terms, probable amounts, and timing of such loans during the construction period need to be considered. Assurance that the funds would be forthcoming is essential even before consideration is given to bidding. If this is not done, there is a risk in having to pay exorbitant rates to the party willing to advance funds in an emergency situation.

Payment terms, the withholding of reserves, and the reduction of reserves upon completion of half the work should be studied with regard to their influence on job financing. Monthly payments by the owner should be arranged to set up a cash flow schedule to take care of subcontractors but to minimize the payments which the contractor himself has to carry.

Performance and payment bonds for a private job such as we are considering ordinarily would not be required. However, if the job is of a public nature with a bond requirement, the contractor should determine in advance whether a bond could be secured. Naturally, there is no point in even bidding the work if there is a possibility of a problem in securing a bond after the necessary trade investigation.

GENERAL CONSIDERATIONS

The proposal and contract forms for the agreement with the owner may require examination by an attorney to avoid overlooking any especially onerous clauses. Special provisions affecting subcontractors should be extracted and included in the bidding information together with a reference to the subcontract form itself, which should be available for examination.

Discussions with an insurance broker regarding limits, hold harmless clauses, and special coverage, such as XCU (for explosion, collapse, and damage to underground utilities), may be advisable for hazardous work. Care, custody, and control provisions for adjacent properties and contract floater policies for special risks in handling and transportation are all possibilities needing appraisal.

Information gathered by various individuals should be turned over to the estimator, who will make it available to those responsible for transmitting to subcontractors or those on the staff requiring the data.

Some variations with respect to submission of proposals should not be overlooked. For instance, a proposal on public work, which was to be accompanied by a certified check for surety, could be rejected as unresponsive if a surety bond were submitted instead of a certified check.[10]

Bidders generally post a certified check or a bid bond. The bid bond assures that if a contract is awarded to the contractor, he will, within a specified term, sign the contract and furnish the specified performance bond.

It should be noted that most surety companies authorize bid bonds after the performance bond on a particular contract has been underwritten and approved.   For this reason, contractors are cautioned against depositing a certified check with a bid unless there are assurances that the performance bond on that particular contract will be underwritten and approved.

BONUS AND PENALTY CLAUSES— LIQUIDATED DAMAGES

Some projects, usually but not necessarily of a public nature, may be of such importance that a bonus will be paid if the work is finished and accepted before its scheduled completion date, and a penalty will be assessed (liquidated damages) for each day the job remains uncompleted.

Extensions of time for delays beyond the control of the contractor and/or subcontractor will be granted to the extent specified in the agreement.   A general contractor performing under such an agreement would normally arrange with his major subcontractors for a proportionate amount of bonus or penalty.

If a bidding document includes bonus or penalty clauses, it may be reason to forgo estimating the work, because only experienced contractors and subcontractors should become involved.   The contracting authority, in endeavoring to be fair, can be expected to grant extensions only for completely justifiable reasons.   In an Oklahoma case on a government contract, an appeal was made by the contractor to the General Services Administration Board of Contract Appeals to reduce the liquidated damages assessed for delay caused by (1) failure of two utility companies to finish their work on schedule and (2) shortage of masons.

The Board stated that the utility work, not part of the prime contract, did affect completion of the contractor's work and that liquidated damages should not be charged for this time.

The Board recognized the contractors exhaustive but unsuccessful efforts to recruit both union *and* nonunion masons but concluded that delay due to a labor shortage is not exclusive grounds to grant relief, unless the government by some act caused the labor shortage.[11]

At best, in some cases, extended appeals may secure relief for extenuating circumstances.   At the worst, some of the best-qualified witnesses may no longer be available to testify, and the appeal may have to be abandoned after expensive litigation has occurred.[12]

If any liquidated damage clause is used, it should include a clear statement that the subcontractor agrees that he can complete the work in the number of days specified; that he guarantees to complete the work in the time stated; and that for each and every day beyond the days specified in the subcontract that are required for the completion of the work, the subcontractor shall pay the contractor the sum of $_____ as liquidated damages.

The word "penalty" should not be used, for there should not be a penalty clause without a corresponding "bonus" clause.

NOTES—Chapter 1

1. Robert Snow Means Company, Inc., "Building Construction Cost Data," Duxbury, Mass., 1973, p. 194.
2. Reuben Samuels, at ASCE-Metropolitan Section Seminar on Excavation and Foundation Bidding, Apr. 24, 1972.
3. Marvin Gates, "Bidding Contingencies and Probabilities," *Journal of the Construction Division: Proceedings of the ASCE,* November 1971.
4. *Anthony P. Miller, Inc. v. Wilmington Housing Authority,* 179 F. Supp. 199, 184 F. Supp. 273.
5. *Shapiro v. Driscoll Co.,* 266 App. Div. 260, 42 N.Y.S.2d 94; *aff'd,* 292 N.Y. 519, 54 N.E.2d 205 (N.Y. 1944).
6. Alan Greenspan, "The Escalation of Wages in Construction," The Associated General Contractors of America, Washington, D.C., June 1970.
7. *Heifetz Metal Crafts, Inc. v. Peter Kiewit Sons' Company,* 264 Fed.2d 435 (1959).
8. *Uniform Construction Index,* The Construction Specifications Institute, Washington, D.C., 1972.
9. James J. O'Brien, P. E., *CPM in Construction Management: Scheduling by the Critical Path Method,* 2d ed., McGraw-Hill, New York, 1971.
10. *Stage v. Whitehouse,* 48 Misc.2d 703.
11. Appeal of Clyde Burton & Son Construction Co., Inc., GSA BCA #3227, 71-2 BCA 9152 (1971).
12. A governmental form used during World War II was similar to the following: If the subcontractor's right to proceed is so terminated, the contractor may take possession of and utilize in completing the work such materials, appliances and plant as may be on the site of the work and necessary therefor.

    If the contractor does not terminate the right of the subcontractor to proceed, the subcontractor shall continue the work, in which event the actual damages for the delay will be impossible to determine and in lieu thereof, the subcontractor shall pay to the contractor as fixed, agreed and liquidated damages for each calendar day of delay until the work is completed, the sum of $_____, and the subcontractor and his sureties shall be liable for the amount thereof; if the date of completion cannot be determined by and between the contractor and subcontractor hereunder, the completion date shall be determined as set forth in Article _____ of the subcontract relative to settlement of disputes. Provided, however, that the right of the subcontractor shall not be terminated or the subcontractor charged with liquidated damages because of any delays in the completion of the work due to unforeseeable causes beyond the control and without the fault or negligence of the subcontractor, including, but not being restricted to, acts of God, or the public enemy, acts of the Government, acts of another contractor in the performance of a contract with the Government, fires, floods, epidemics, quarantine, restrictions, strikes, freight embargoes and unusually severe weather or delays of other subcontractors due to such clauses, if the subcontractor shall within ten (10) days from the beginning of any such delay notify the contractor in writing of the causes of delay, who shall ascertain the facts and the extent of the delay and extend the time for completing the work when in his judgment the findings of fact justify such an extension, and his findings of fact shall be subject to the provisions set forth in Article _____of this subcontract. The contractor may deduct the amount of accrued liquidated damages from any monies due or to become due to the subcontractor.

CHAPTER TWO

Bid Analysis, Estimate Preparation, and Submission of a Lump-Sum Proposal

Having received most of the bids and listed them on the tabulation forms, we are now ready to study them and note the qualifications or omissions. These will be appraised and adjustments made if necessary to complete the picture for each trade. The omissions may reflect the subcontract bidders' inability to complete all phases of the bid preparation in the allotted time or may indicate a desire to be called in to discuss the bid.

All bidders want to know the relationship of their appraisals of the job to those of others. After some study of the bids, the contractor is usually in a position to pass along some information to inquiring bidders without disclosing information which would be detrimental to other bidders. It might be stated that the bid is "in the ball park" or completely out of the picture. By these discussions, it is possible for a bidder to get a better perspective for future bidding and possibly to revise the bid previously submitted for the current project.

After considering the various bidders' qualifications, the closeness of their bids, and the spread between the first and second bids, and after sizing up local building conditions, especially the probable supply of labor, the general contractor is able to decide how much of a contingency or discount should be applied to finalize the price to be submitted for the prospective contract.

If it is important to win the job, the proposal submitted by the general contractor should be "tight" (low), while if a contractor expects to be overloaded with work, the price for the project will probably be high.

To quote a fair overall price to a prospective client, the contractor is always faced with the problem of summarizing the subcontractors' bids realistically. To use some bids as submitted, even assuming that all questions apparently were resolved, is not always the best policy. Some bids, for example, may be ridiculously low because of computational errors or deliberate omissions, while other low bids, when studied carefully, may be found to be too high. There are those to whom the job has become uninteresting; or there are some unattractive features which have become apparent to the bidders. Some subcontractors may be inclined to call the estimator of another subcontractor and add or deduct up to 10 percent of the competitor's bid for their own estimate.

Bids that do not specifically respond to the bidding inquiry and confirm its specific requests should be verified carefully. For instance, supplementary information issued during the bidding period should be checked regarding its inclusion in the bids.

Occasionally, a bidder will intentionally omit the reference to some items not specified but which an examination of the site would disclose and which, should a contract be awarded, would be grounds for additional reimbursement to the subcontractors on the basis of "changed conditions."

On the other hand, the Agricultural Board of Contract Appeals, with respect to reimbursement for 35,000 tons of fill needed for a haul road, ¼ mile of which became a lake after rain, ruled that no additional compensation was warranted since the ground conditions were essentially the same as when the site on the dry lake bed was available for inspection. The contract had provided that no payment was to be made for "borrow" used to build the road.[1,*]

In this connection, a summary of the four categories of changes requiring an analysis before an accurate determination of responsibility can be made is given by George F. Sowers, Professor of Civil Engineering, Georgia Institute of Technology, as follows:

1. Changes that are not foreseeable by a reasonable subcontractor

2. Foreseeable changes, either produced by the construction method or inherent in the environment

3. Pseudo changes, foreseeable changes which are preventable and which are caused by faulty construction methods

* Superior numbers refer to Notes at end of chapter.

4. Mistaken changes or problems that are wrongly attributed to changes in site conditions[2]

Subcontract bids for a project cannot always be expected to be submitted exactly in accordance with the drawings and specifications and the bidding information. Life would be much easier for the purchasing man if they were, and the low bidders then would be the logical subcontractors to deal with. In practice, the qualified bids require a very careful analysis and open up the possibility of different approaches to the final selection of a subcontractor and the price that can be negotiated.

It is worth remembering that some building specifications may be so extremely restrictive that subcontractors will have little leeway in estimating. Other specifications, by giving more latitude, can encourage and increase competition among subcontractors while still affording a first-class result for the owner.

Let's take some of the bids received, adjust them to a comparable base, and then determine the amount that should be used on the estimate summary sheet (Fig. 5). There will be other points to be investigated, but these can be deferred until more time is available if and when the general contract is secured.

In the examples that follow, it is assumed that the items excluded by the bidders are properly part of the specified work and should have been included in the bid, and that the alternates quoted, for whatever reason, should not have been included in the base bid. Unit prices, indicated as plus or minus, represent figures to apply to additions or deductions, respectively, from the contract for the work.

The qualifications noted for the various bids, under Remarks, are intended to illustrate a range of deviations from normal specification requirements. Experience in the area and the contractor's own knowledge of their feasibility will determine to what extent they offer an opportunity for bargaining.

For illustrative purposes, additional assumptions occasionally are given in the analysis of the bids to justify the reasoning in arriving at the amounts of the discounts, if any, or the reliability of the various bidders. An asterisk indicates the bid selected.

EXAMPLES OF BIDS

Demolition

(a) OUTLINE SPECIFICATIONS: Demolish two three-story and basement masonry and frame houses. Combustible material and scrap to be removed from site; noncombustible debris may be left in cellar. Obtain all permits.

(b) BIDS

Remarks

A	$4,750	
	+ 50 (bond)	
	$4,800*	
B	No bid	
C	$8,000	

(c) ANAYLSIS: While a third bid would have been helpful to reinforce the analysis, the contractor recently awarded a demolition contract for a similar house in another area for $2,000. Even though the amount involved is comparatively small, it is advisable to recognize that the low bidder is unknown to the contractor, that he is working in an area away from the contractor's home base, and that there is some valuable material on the premises to be salvaged. Demolition bids involve labor and trucking against which salvage is credited.

In this case, a performance bond, adding 1 percent to the bid amount, would be warranted to avoid having the subcontractor start work, remove the valuable salvage, and leave the contractor to worry about having to perform the balance of the demolition with his own men.

DECISION: Use $4,800.

Excavation

(a) OUTLINE SPECIFICATIONS: Two core borings at the site of the proposed building indicate rock approximately 8 feet below sidewalk grade with ground water 5 feet below sidewalk. Excavate rock in elevator pit. Underpin and brace wall of adjoining building. Provide sheeting and bracing of earth banks to retain curbs and roadways. Include all permits.

(b) BIDS

Remarks

A	$20,000*		Telephone bid
B	$30,000		
	−1,000	(mechanical trenches)	HVAC includes excavation for
	$29,000		mechanical trenches
C	$32,000		
D	$40,000		
	−5,000	(paving)	Site work includes paving
	$35,000		

	Time (weeks)	Unit prices/cu yd Earth	Rock
A	9	$4.50/4	$17/15
B	7	$4.75/4	$18/15
C	10	$5.00/4	$17/16
D	12	$6.00/5	$20/18

(c) ANALYSIS: The bids have been adjusted by deductions to make them comparable with the scope of the work as defined in the specifications.

Using the contractor's takeoff quantities for 1,500 cubic yards of earth and 500 cubic yards of rock with the low bidder's deductive units of $4 and $15 per cubic yard, respectively, the base estimate amounts to $13,500, to which will have to be added other costs as noted:

Earth	1,500 cu yd @ $4	$ 6,000
Rock	500 cu yd @ $15	7,500
Underpinning		5,000
Bank sheeting, bracing		5,000
Miscellaneous items (elevator pit,		
ramp, watchman, pumping)		5,000
Contractor's approximate estimate		$28,500

While there is no reason to suspect that A's telephone bid of $20,000 was received inaccurately, the above analysis and the relationship of the other bids seem to indicate the possibility of an error. We will assume that bidder A cannot be recontacted and also that the other general contractors may use his $20,000 bid.

DECISION: Use $20,000, but set up a $5,000 contingency in the event the bidder ultimately would not accept a contract at $20,000 if it were offered to him.

There are other possibilities here worth considering, such as:

1. If it had been possible to reach bidder A with a suggestion that he recheck his estimate, an error might have been found and the bid revised. In this case, the bidder would have been expected to notify the other general contractors accordingly.

2. After discussion, bidder A might have agreed to hold to the $20,000 figure, recognizing its low feature but wanting especially to secure this job because of idle equipment he would be able to put to use. The bid would be considered exclusive to the contractor who, if he secured the contract, agreed that he would use A as his excavation subcontractor. Other contractors had been given a price of $26,000.

Structural Steel

(a) OUTLINE SPECIFICATIONS: Type A-36 steel; H section columns; 12-inch beams, 18-inch girders with two 10-inch filler beams per bay; high-strength bolting where specified; attached window lintels. Apply field coat paint where steel is not encased in concrete. Inspection by owner's engineer.

(b) BIDS

		Approximate tonnage	Delivery (weeks)	Remarks
A	$63,000	150	20	
	+1,000			Excludes field paint
	$64,000/150 = $427/ton			
B	$61,000	135		
	+1,000			Excludes field paint
	$62,000*/135 = $460/ton		21	
C	$73,000	160	22	
	+1,000			Excludes field paint
	$74,000/160			
D	$70,000		16	
	+1,000			Excludes field paint
	$71,000			
E	$73,000	150	13	Warehouse delivery
	+1,000			Excludes field paint
	$74,000/150			Price good 14 days
F	$50,000			No erection included

(c) ANALYSIS: The first question to be answered here is whether there would be an advantage to be gained by using E's time of 13 weeks for warehouse delivery rather than mill delivery. Better timing is always advantageous, especially when it involves materials on hand (which can be verified). Based on a discussion with the excavation bidders, it is questionable whether rock excavation, underpinning, and foundation work could be advanced to meet the early steel delivery without excessive overtime. In fact, the 16-week delivery time of bidder D would still be less than the anticipated excavation and foundation progress. Furthermore, the 14-day option period, practically speaking, makes E's bid unacceptable for the current estimate, since it is highly unlikely that a contract would be awarded that quickly.

All bidders excluded field painting, which could be awarded either to a structural-steel painter or to the job's painting subcontractor. One of these subcontractors quoted $10 per ton for one coat. Using this price for 100 tons (some unpainted), an allowance of $1,000 on all bids should cover this item.

The choice narrows down to:

A $64,000/150 tons/20 weeks = $427/ton in place
B $62,000/135 tons/21 weeks = $460/ton in place

As these are lump-sum proposals, the tonnage figures are only rough approximations (some bidders might purposely show a higher tonnage to impress the buyer). Nevertheless the bids indicate average prices of $427 and $460 per ton, which both bidders probably realized when submitting.

Using B's tonnage and A's average unit price (which is admittedly risky), the estimated price would be 135 × $427, or $57,645. Note, however, that the average of the four tonnage approximations is 149 tons so that B's number of 135 tons appears low.

Bidder F quoted only on delivering the steel. It is assumed that there is only one available erection subcontractor in the area and that the contractor has had poor experience in awarding separate erection subcontracts to handle steel purchased f.o.b. job site. In some localities, it might be possible to rent equipment and hire ironworkers for the job. In this case, a contract to furnish *and* erect the steel is highly desirable to avoid coordination problems which could occur otherwise.

The contractor's interest in securing the contract will determine whether to use the $64,000 bid of A for 20 weeks or B's bid of $62,000 for 21 weeks. Should the contract be secured, the contractor undoubtedly would find B interested in the job at a price of $61,000 but now including the allowance for field paint. B would probably also agree to perform in 20 weeks, making it possible to narrow the comparison between A and B to a price basis, without time factors affecting it.

DECISION: Use $62,000 with $2,000 discount.

Larger steel subcontracts are frequently awarded on the basis of an all-inclusive lump-sum unit price per ton based on the provisions of the American Institute of Steel Construction code for building construction, with estimates of the tonnage furnished.

The advantage here is that it permits an earlier receipt of bids and determination of the fabricator since the estimating time will be a week or so less than in the preparation of a lump-sum estimate. The early award of a steel subcontract can help greatly in the completion of any building.

The drawings prepared by the structural engineer are presumed to be complete enough for estimating purposes since they will indicate the types of materials, connections, and general categories of the work and permit initial mill ordering of material. The final costs depend on the computed weights as determined from the approved shop drawings in accordance with the American Institute of Steel Construction code times the subcontract unit prices, and the contract with the owner will be adjusted proportionately.

If the category of work changes and becomes more expensive, the fabricator will normally expect to be reimbursed. Unit prices for additions and deductions are usually quoted for changes with framing similar to the basic job.

On these larger jobs it is normal practice to recognize the "timing" when the change will take place as the unit prices are agreed upon. For instance, a change of sizing prior to the preparation of shop drawings, mill orders, etc., should be less expensive to handle than if the work were actually underway, in which case the unit prices would require adjustment or would not apply.

Most steel-erection departments base their estimates on the number of pieces handled. Consequently, an increase in weight of a particular member would normally result only in an extra charge for furnishing the extra steel.

While there are accepted engineering standards for steel construction and all work would be expected to conform to the structural engineer's specifications, it should be recognized that fabricators quoting on a unit price per ton for the job are not under the same compulsion to minimize the tonnage furnished as they would be in a lump-sum contract.

Concrete

(a) OUTLINE SPECIFICATIONS: 2,000 psi concrete for footings, foundation walls, slab on grade; 3,750 psi lightweight reinforced slabs for floors and roof. Pumping as required while work is in progress. Subsoil drainage. Metallic waterproofing for basement floor and walls to 6-foot height. Cement finish floors except where tile or terrazzo is shown. Cement finish on stair treads and platforms. Include mix design. Inspection by owner's engineer.

(b) BIDS

		Unit prices (2,000 psi)	Remarks
A	$87,000	$125/110	
	+1,000 (pumping)		Excludes pumping
	$88,000*		
B	$103,000	$120/105	
	−1,000 (concrete bases)		Bases specified under HVAC
	$102,000		
C	$110,000	$130/95	
D	No bid		± 700 cu yd

(c) ANALYSIS: It is important when considering bids that appear low in relation to others that the bidder be given an opportunity to recheck his estimate. At the same time, the responsibility for the quantities and the pricing should be the bidder's. The contractor should be careful not to find himself in the position of having needlessly guaranteed quantities which could vary from those developed more accurately later after additional examination of the drawings and specifications.

In this particular case, bidder A has a good reputation and is completely satisfied with his figure after a suggestion that he check his bid and quantities.   He has excluded the specified pumping which the contractor estimates would be worth $1,000.   There are advantages in having the concrete subcontractor responsible for this service as specified, rather than having the general contractor become involved with extra claims for failure to provide pumps needed to permit other trades to work.   Under abnormal conditions, such as severe storms, pumps are sometimes hard to come by, and the contractor might well avoid agreeing to take on something that the concrete subcontractor would accept initially without much urging (unless the contractor's own equipment is available and could be furnished on very short notice).

B's bid has been adjusted to exclude the concrete equipment bases which were specified under heating, ventilating, and air conditioning.

Some subcontractors are hesitant about revealing their takeoff quantities of materials.   When subcontractors offer quantity information, it is only fair that the general contractor advise them if their figures differ greatly from their competitors' or his own.   Most subcontractors realize that the discussions are intended primarily to avoid any gross errors in estimating the job, which is of mutual benefit to both parties.

A decidedly low bid does not automatically indicate error.   Perhaps the low bidder is the only subcontractor who is estimating the job properly.

It sometimes happens that bidders are unable to complete their proposals but are quite willing to pass along such quantity information or other data as they have had a chance to assemble.   In the case of D, using his estimate of 700 cubic yards and A's deduct unit price of $110 per cubic yard, the extension of $77,000 plus subsoil drainage and metallic waterproofing checks A's bid reasonably well.   It must be recognized that an average unit price for concrete on most jobs should be used sparingly, as there may be several different types of concrete, which affect material as well as placing costs because of the different categories of work.

After A's bid has been adjusted to include pumping, it seems reasonable to expect that A would be willing to take on the pumping obligation in the "buy out."

DECISION: Use $88,000 with $1,000 discount.

Masonry

(a) OUTLINE SPECIFICATIONS: Red face brick for exterior walls with 8-inch lightweight concrete-block backup as required by local code.   Stairways 8-inch concrete block as per code.   Alternate #1: Provide 3-inch hollow cinder-block partitions to underside of ceiling slabs, as detailed.

(b) BIDS

		Remarks
A	$63,000	All bids exclude winter-weather
	+5,000	protection, patching, and
	$68,000*	hoist charges
B	$67,000	
	+5,000	
	$72,000	
C	$66,000	
	+5,000	
	$71,000	
D	$70,000	
	+5,000	
	$75,000	

(c) ANALYSIS: All bids excluded winter-weather protection and expenses such as heating mortar or supplying canvas to permit work to proceed in colder areas. The contractor would do well to have the subcontractor responsible for his own protection to avoid later arguments. Patching, usually caused by other trades, could be handled as an extra to the masonry subcontractor with costs backcharged to the subcontractors responsible. To provide for these exclusions, $5,000 has been added to all bids by the contractor.

The period for performing the masonry work is such that the contractor could spare the use of the hoist and save the expense of a separate hoist installed by the masonry subcontractor. If this arrangement is followed, the masonry subcontractor should carry the hoisting engineer on his own payroll together with the laborers servicing the material handling. In the event of a failure on the part of the engineer, he would be serving as an employee of the subcontractor, and his actions would normally not attach responsibility to the contractor.

The inclusion of any hoisting charges in the subcontractor's estimate, for an installation furnished by the contractor, increases the cost of the work and consequently the fee or profit which the subcontractor would normally include. There is a fine line involving hoist charges in that free usage may save some fee but result in considerably less control of the hoisting operation on a multistory building or a job where minimum hoisting facilities are available.

A charge of $30 per hour for straight-time or overtime use of the hoist is commonly made in some metropolitan areas, but practices vary because of the many factors affecting this service. The contractor's total rental collections for hoist usage will be recognized when computing the overhead, or "back-page," items as a credit to such costs.

The $3,000 discount would represent a conservative amount which might be expected in the buy out of a subcontract of this size, with the range of bids indicated.

DECISION: Use $68,000 with $3,000 discount.

The bidders quoted on Alternate #1 (Interior partitions) as noted below as well as on Alternates #2 and #3 for setting sills and caulking. Bids of A and C will help in analyzing the stonework picture.

Alternates	#1 Interior partitions	#2 Set sills	#3 Caulking
A	+$3,000 (Error?)	+$600	
B	+$8,500		+$600
C	+$7,000	+$400	+$1,200
D	+$7,500*		
Average B/C/D	+$7,666		

DECISION: Use $7,500 for Alternate #1.

Stonework

(a) OUTLINE SPECIFICATIONS: Domestic granite, 3 inches thick, with 12-inch-high base course and granite at three entrances.   Stainless steel anchors.   4-inch limestone window sills.

(b) BIDS

			Remarks
A	Furnish and set granite	$5,500	
	Furnish limestone	+700	
		$6,200	
	Unload granite	+300	Contractor's estimate
	Set limestone	+600	Mason's bid
		$7,100	
B	Furnish and set granite; set limestone	$5,900	
	Furnish limestone	+700	
		$6,600	
	Supply stainless steel anchors	+200	Contractor's estimate
		$6,800	
C	Furnish and set granite	$7,000	
	Furnish limestone	+800	
		$7,800	
	Set limestone	+600	Mason's bid
		$8,400	
D	Furnish limestone	$ 600	
E	Furnish and set granite	$5,400	
		$6,000	
	Set limestone	+600	Mason's bid
		$6,600*	

(c) ANALYSIS: It is usually advantageous to combine all stonework in one subcontract, especially if stone setting is involved.   The cutting and fitting as well as damage claims then become less controversial.

In this building, the setting of limestone slip sills bears no relation to the granite setting at base courses and entrances. The limestone sills could be set (in terms of jurisdiction) by the masonry subcontractor. This would permit the most advantageous purchase of sills through D and the furnishing and installation of granite by E. While there are two bids from masonry subcontractors in the amounts of $600 and $400, the masonry subcontractor who was tentatively selected for the trade was using a $600 bid on this limestone. This should be used at this stage since we have no assurance, especially with three parties involved, what the total stone package could cost. If the granite subcontractor is still on the job when the time comes to set limestone, he might take on this setting also.

The closeness of the bids and the number of items involved still offer a possibility of savings which conservatively was set at $200.

DECISION: Use $6,600 with $200 discount.

Windows and Storefronts

(a) OUTLINE SPECIFICATIONS: Center-pivoted aluminum windows; Duranodic finish.   Aluminum display windows; entrance doors.

(b) BIDS

				Remarks
A	Window	$15,500		
	Storefront	+21,000		
		$36,500		
		+1,000	(protection)	Excludes protection; contractor
		$37,500		estimates $1,000
B	Window	$20,000		
	Storefront	+21,000		
		$41,000		
C	Window	$14,500*		
	Storefront	+20,500		
		$35,000		
D	Storefront	$19,000*		

(c) ANALYSIS: The window and storefront work was specified in one section of the specifications.   Bidder A is unwilling to take these portions separately, whereas bidder C would agree to separate.   The bids are close, the job is not large, and there are advantages in having one subcontractor do both portions of the work which have certain similarities.   A combination of C's window bid and D's storefront bid represents the current low estimates.

Normally when securing proposals where one subcontractor may only quote on a portion of the work, it is well to reserve the right in the invitation to bid to award the work separately.   In this case it was not done, and bidder A has every right to expect that his proposal would only be considered in its entirety. The closeness of the C and D combination bid or bid C would indicate that a $1,000 discount would be conservative.

DECISION: Use $33,500 with $1,000 discount.

Glazing

(a) OUTLINE SPECIFICATIONS: Clear polished plate glass for all windows, storefronts, and entrance doors. Thickness of all glass to conform to manufacturers' standards for size lights shown.

(b) BIDS

Remarks

A	$10,000*
B	$12,500
C	$13,500

(c) ANALYSIS: The spread between first and second bid is such that a discount is not recommended.

DECISION: Use $10,000.

Miscellaneous Iron

(a) OUTLINE SPECIFICATIONS: Steel stair construction with channel or plate stringers; concrete-filled treads and poured-concrete landings; pipe rails as detailed. Trench pit frames and covers; ladders; rough bucks for elevator door openings; ceiling-hung toilet partitions.

(b) BIDS

	Toilet parti- tions	*Steel stairs*	*Built-in items*	*Other miscellaneous*	*Total*
A	$1,500	$8,100	$2,000	$3,600	$15,200
B		$7,200	$1,500 (incomplete)	$2,900 (incomplete)	$11,600
C	$1,100	$8,200	$2,400	$4,300	$16,000
D	$1,300		$ 600 (incomplete)	$1,100 (incomplete)	$ 3,000
Low bids	$1,100	$7,200	$2,000	$3,600	$13,900

(c) ANALYSIS:

Remarks

A	$15,200	
B	$11,600*	Incomplete
C	$16,000	
D	$ 3,000*	Incomplete

Frequently specification sections, such as miscellaneous iron, include somewhat unrelated items. Toilet partitions, rolling steel doors, special metal accessories, and others may be lumped into one section. Some bidders will quote a lump sum for all the work, while others will quote only on certain items. Normally a complete breakdown of the bids would not be requested, but the invitation to bid should stipulate that partial bids would be considered.

Although it is usually advisable (and certainly most subcontractors prefer) that the miscellaneous iron requirements should not be split into several subcontracts, some specialty subcontractors handling toilet or wire-mesh partitions, stairs, or shutter doors may be accustomed to quoting to the contractor directly and favorably, permitting the work to be handled independently of other miscellaneous iron.

If we use the prices assumed for the various subcontractors, it is apparent that by combining B's and D's incomplete bids, we have a combination of $14,600 for a complete package compared to A's and C's higher bids. The total of all the low bids is still only $13,900, which could serve as a target for future negotiation.

It should be verified that the low-bid masonry and concrete subcontractors have included provision for building in items furnished f.o.b. job site.

DECISION: Use $14,600 with $600 discount.

Ornamental Metal

(a) OUTLINE SPECIFICATIONS: Aluminum and glass partitions at elevator lobbies (quote separate price but include in lump-sum bid); aluminum door saddles, mail chute, and letter box in lobby; directory board; exterior signs on building.

(b) BIDS

	Aluminum and glass partitions	All other	Combined	Remarks
A	$ 5,000	$10,000	$15,000	
			+500	Cleaning; contractor's estimate
			$15,500*	
B	$10,700	$ 5,200	$15,900	
C	$11,100	$ 5,400	$16,500	
D	$ 7,500	$12,500	$20,000	
E		$11,000		Incomplete

(c) ANALYSIS: The ornamental-metal specification was divided into two parts with the intention of securing a separate bid on aluminum and glass partitions which was to be included in the total bid.

Some of the pricing was obviously incorrect. Telephone calls to B confirmed that he had reversed the amounts when he submitted his combination bid, and that it should have read $5,200 for the separate price and $10,700 for the balance.

Bidder C did not return a telephone call and was apparently one of a group who apportioned work among themselves and who had received his figure elsewhere without estimating the work personally.[3]

The contractor estimated that the cleaning which A had excluded is worth $500 and verified this amount when talking with B. There is apparently confusion in this trade and discounting could be risky, although it would be reasonable to expect A to agree to perform the cleaning when closing out a contract.

DECISION: Use $15,500 with $500 discount.

Waterproofing, Dampproofing, and Caulking

(a) OUTLINE SPECIFICATIONS: Spandrel at all window and door heads as detailed; one brushed-on coat of dampproofing on interior face of exterior walls; caulk against masonry at all windows.

(b) BIDS

			Remarks
A	Waterproofing	$3,500	Not as specified
	Dampproofing	+2,000	
	Caulking	+1,000	
		$6,500	
B	Waterproofing	$4,500	
	Dampproofing	+2,100	
	Caulking	+1,100	
		$7,700*	
C	Waterproofing	$4,500	
	Dampproofing	+2,100	
	Caulking		
		$6,600 (incomplete)	No return trips
D	Waterproofing	$5,800	
	Dampproofing	+2,300	
	Caulking	+1,200	
		$9,300	

(c) ANALYSIS: A's bid was based on an unspecified material. The use of this figure would be reasonable only if the contractor had positive knowledge of its equivalency and acceptance by the architect. Savings accruing to subcontractors from other-than-specified products are usually minor but can cause excessive losses if replacement becomes necessary.

The restriction by C as to a limited number of trips to the job is a nuisance and tends to inconvenience job operations. Where many trips and comebacks are necessary, especially on isolated jobs, a price for each trip over some base number should be established and arrangements made to compensate. Then it becomes the responsibility of the contractor's superintendent to carefully coordinate the need for the return trips.

It is usually good practice on small contracts of this nature to combine the work of several trades in one agreement (frequently the mechanics are in the same union), not only because of more efficient field operations but also because of fewer problems in having the work performed when needed.

Note that other subcontractors, such as those performing masonry or window work, may also perform caulking. Such a combination of masonry and caulking or window work and caulking might be used to advantage in fixing responsibility to prevent leakage. Bids from these subcontractors may be helpful in determining the final allocation of the work as well as the best price.

DECISION: Use $7,700 with $200 discount.

Hollow-Metal Work

(a) OUTLINE SPECIFICATIONS: 18-gauge doors, flush design, 1¾ inches thick; 16-gauge bucks throughout with combination buck, frame, and trim; baked prime finish.

(b) BIDS

	f.o.b.	Erection	Combined	Remarks
A	$4,000	$2,000	$6,000*	
B	$4,200	$2,100	$6,300	
C	$3,900	$2,200	$6,100	
D		$1,800		
E		$1,960		Contractor's estimate

(c) ANALYSIS: The bids on this work are close and well confirmed by the contractor's estimate on the erection work as follows:

Erect 40 frames @ $10 each = $ 400
Erect 40 doors @ $39 each = +1,560
$1,960

or using

C	f.o.b. price	$3,900
E	Erection	+1,960
E	Field measure	+100
		$5,960

At this stage it seems advisable to use the low bid on the work erected in place and defer the decision on whether to subcontract the erection or to purchase the materials and handle the erection with the contractor's own men. There are advantages in having odd jobs for the carpenter to keep busy with; frequently it is hard to find an erection subcontractor for those jobs which do not involve continuous periods of work.

DECISION: Use $6,000.

Finish Hardware

(a) OUTLINE SPECIFICATIONS: Corbin #300 series or Russwin #500 series locks; 1½ pairs ball-bearing butts per door; exposed members of all hardware to be stainless steel except closers, and butts to have prime coat for painting, all per schedule.

(b) BIDS

Remarks

A	$2,500
	+100 (tax)
	$2,600*
B	$2,700
	+108 (tax)
	$2,810
C	$2,800
	+112 (tax)
	$2,910

(c) ANALYSIS: Some trades, such as finish hardware, usually involving comparatively small costs and little chance for savings in the purchase, nevertheless can exert a considerable effect on the progress of the job.

A conscientious hardware dealer who orders the "right" hardware, gets the templates to the other subcontractors when needed, and follows through for delivery from the manufacturers and/or corrections, can be of great benefit to the job. Consequently, it is debatable whether much effort should be given with such trades to create substantial savings in buying. A "target" might be to determine if the hardware vendor would willingly absorb the sales tax in his price.

The need for service is just as great on the larger buildings, but the dollars at stake warrant great care in the negotiation and establishment of reasonable unit prices.

DECISION: Use $2,600.

Roofing, Sheet Metal, and Membrane

(a) OUTLINE SPECIFICATIONS: Rigid insulation—1½-inch fiberglass on all surfaces; 2-ply vapor seal under insulation with 4-ply felt and slag finish. Base and counter flashing to be .032 mill-finish–0-temper aluminum, coated both sides with Bitumastic enamel. 2-ply membrane waterproofing under all toilet and slop sink room floors.

(b) BIDS

Remarks

A	$5,200	
	+300 (hoisting)	Excludes hoisting; contractor's estimate
	$5,500*	
B	$5,700	
	+100 (additional trips)	Two additional trips; contractor's
	$5,800	estimate
C	$9,000	Not approved applicator
D	$10,500	

(c) ANALYSIS: The subcontractor for roofing should always be an applicator approved by the manufacturer whose products could be used. Hoisting materials is part of the work. There is less risk from the contractor's standpoint if subcontractor remains responsible for raising his materials by some means, not necessarily using the contractor's hoist.

Extra claims for return trips to apply membrane are bothersome and should not be necessary if the work is carefully coordinated by contractor and subcontractor.

Bids are close; either A or B can be expected to compete for the job.

DECISION: Use $5,500 with $200 discount.

Tile and Terrazzo

(a) OUTLINE SPECIFICATIONS: Toilet rooms to have vitreous tile, coved base, and ceramic-glazed-tile walls full height. Install by "mud" method. Venetian-type terrazzo floors in main entrance and first-floor elevator lobby.

(b) BIDS

	Tile	Terrazzo	Remarks
A	$16,800	$4,000	
B	$15,000	$3,500	
C	$14,500*	$3,000*	
D	$14,000		
E		$3,500	

(c) ANALYSIS: It is common practice to combine these trades in one subcontract, thereby reducing job overhead for the subcontractor. An award to C would be recommended with the expectation that a $1,000 saving could be negotiated.

DECISION: Use $17,500 with $1,000 discount.

Lath, Plaster, and Acoustical

(a) OUTLINE SPECIFICATIONS: Lath chases in masonry; provide hangers and carriers for acoustical ceilings. Exterior walls of second to fourth floors and interior columns to be plastered; three coats on metal lath and two coats on masonry; last coat hard white finish. Include scratch coat for ceramic tile. Ceilings in public areas, elevator lobbies, and rentable areas second to fourth floors to have suspended concealed-runner-type construction, 4 feet apart. Runners to support standard recessed modular fluorescent lighting fixtures, mineral-fiber perforated tile, and concealed spline joints; sound-reduction coefficient of .70.

(b) BIDS

	Lath	Plaster	Acoustic	Remarks
A	$6,500			
B	$8,000	$10,000	$14,000	$10,000 excludes patching; contractor's estimate for patching $500
				$14,000 includes hangers and carriers
		+500		
		$10,500		
C		$12,000	$14,000	$14,000 excludes hangers and carriers
D	$7,500	$11,500		
E			$13,500	Includes hangers and carriers
F		$10,000		

(c) ANALYSIS: The best prices for the work are:

A	Lath	$6,500
F	Plaster	+10,000
E	Acoustic	+13,500
		$30,000

Some subcontractors such as B perform work of all trades. This would be advantageous provided a major price reduction could be worked out. A lath-and-plaster combination or an acoustic contract including hangers and carriers for this work are possibilities, but no more than two subcontractors are recommended for the three trades. The $30,000 target should generate a saving of $2,500 and seems reasonable based on the assumed bids.

DECISION: Use $32,500 with $2,500 discount.

Carpentry and Millwork

(a) OUTLINE SPECIFICATIONS: Fireproof paneling in building owner's office, fourth floor; bookshelving to be shop-primed. Special hardware to be supplied under Finish hardware section.

(b) BIDS

		Remarks
A	$6,500	
	+200	Priming excluded; contractor's estimate
	$6,700	
B	$7,000	
	−300	Hardware included
	$6,700	
C	$10,500	

(c) ANALYSIS: With two identical bids it could be expected that some discount can be obtained. The $200 figure seems conservative.
DECISION: Use $6,700 with $200 discount.

Painting

(a) OUTLINE SPECIFICATIONS: All exterior and interior nonferrous metal to receive two coats of specified paint in addition to shop painting.   All plaster surfaces to receive three coats of paint.
(b) BIDS

		Remarks
A	$11,300	
	−300	Should omit concealed-sprinkler piping
	$11,000*	
B	$11,600	
	−500	Should omit concealed-sprinkler piping
	$11,100	
C	$16,000	
	−1,500	Omit field paint on structural steel
	$14,500	

(c) ANALYSIS: Two bids are very close.   There would be considerable field labor involved, and a greater percentage discount could be expected.
DECISION: Use $11,000 with $1,000 discount.

Resilient Flooring

(a) OUTLINE SPECIFICATIONS: $\frac{1}{8}$-inch vinyl-asbestos floor per floor schedule in colors selected by architect.
(b) BIDS

	Asphalt tile	Rubber base	Combined	Remarks
A	$2,600	$500	$3,100	
			+150 (paper)	
			+320 (wax)	
			$3,570	
B	$3,000	$700	$3,700	
C	$2,700	$600	$3,300*	
D	$3,800	$900	$4,700	

(c) ANALYSIS: Subcontractors frequently exclude paper protection on floors and waxing of tile.   Owner's maintenance staff sometimes performs waxing, and a credit will be obtained from subcontractor.   It is advisable to award subcontract in full accordance with the specifications.   If work is to be omitted, arrangements can be made for credit when the procedure is determined.
DECISION: Use $3,300 with $200 discount.

Elevator

(a) OUTLINE SPECIFICATIONS: Geared traction machine, 54 feet travel, five openings, 150 feet per minute; automatic pushbutton control; 5×7 foot cab; doors, porcelain enamel panels; 2,500 pounds capacity. Baked enamel finish on passenger-elevator doors and frames.

(b) BIDS

		Remarks
A	$39,300*	
B	$41,000	
C	$35,000	Hydraulic
D	$42,000	
E	$50,000	

(c) ANALYSIS: Once a project is out for bids, there is usually not much opportunity for negotiation when it comes to contracting for equipment such as elevators. This is especially true if the equipment is highly specialized and the bidding field is limited.

To verify pricing it is necessary to have a complete description of the equipment proposed and to compare it with earlier purchases for which information is available, such as capacity, speed, travel, number of openings, type of cabs, etc.

For example, a comparison with a similar installation, which had two less openings, could be made by reducing the earlier price by, say, $1,000 for each opening and then updating that cost on the basis of the change in the cost index for the period, with some estimated provision for its later increase. In doing this we assume that in this case the earlier contract was correctly priced and represented a good base to work from.

Competition, the desire of the manufacturer to expand in a new territory, and the closeness of the manufacturing facilities will have an influence on the prices quoted. The number of elevator installations in the surrounding territory and the possibilities for additional maintenance contracts for either elevators, moving stairways, or walkways could affect a particular company's pricing policies.

For estimating purposes, it would be well to use the current low bid.

DECISION: Use $39,300.

If the general contract is secured, it would be worthwhile to explore the possible use of a hydraulic elevator which was quoted on although not specified. If acceptable, this could result in the omission of the penthouse construction, which might be less than the cost of installing the hydraulic plunger below grade and might bring about other savings which the contractor would normally share with the owner on a change of this nature. The knowledge at the time of the buy out that consideration is being given to the change might well affect the final price of bidder A, if it appeared that the job was going elsewhere.

If time permits, a study of possible savings through alternates could increase the owner's consideration of the general contractor's proposal and enhance his opportunities to secure the award.

If the decision is finally made to award the work to A, the possibility still exists that there could be some concessions made, such as a reduced price on replacement elevator cables while car was used for temporary construction hoisting, etc.

Electrical

(a) OUTLINE SPECIFICATIONS: Provide electrical service and switchgear, recessed fluorescent lighting fixtures, switches and outlets for electrical and telephone facilities, all as detailed and specified. Quote Alternate #1 price for temporary light and power and Alternate #2 for alarm system.

(b) BIDS

		Remarks
A	$72,500	
	+5,000 (Alternate #1)	
	$77,500*	
B	$77,000	
	+5,000	Contractor's estimate
	$82,000	
C	$79,000	
	+5,000	Contractor's estimate
	$84,000	
D	$77,500	
	+5,500 (Alternate #1)	
	$83,000	

(c) ANALYSIS: The specifications requested that an alternate price (#1) be quoted for temporary light and power. Both A and D complied. Bidders B and C quoted on a cost-plus-percentage basis. It is usually preferable to have a limit placed on this type of service. B and C should be asked to submit their completed proposals. The extent of the temporary work should be made as definite as possible. In this case, the temporary work, while priced as an alternate, will be awarded with the contract scope and included in the price, as distinguished from other alternates which may or may not be accepted.

For our comparison, we are assuming that B and C would both quote $5,000 for the temporary work, and their bids have been adjusted accordingly.

The bids on Alternate #2 for a special fire alarm system need investigation. If time permits, an effort should be made to see if C's or D's ideas are acceptable.

A	+	$4,000
B	+	$6,000
C	±	0
D	−	$1,000

The bids are comparatively close. Whether the contractor could expect to buy out for much less is problematic, depending largely on the amount of construction work going on and the interest in this particular job. A subcontractor who had performed work on another job for this particular owner might show greater interest than some other electrician.

The wide range of bids on Alternate #2 may require considerable checking. As A's estimate is being considered on the low base bid, it will be well to use his Alternate #2 price of +$4,000 rather than to assume that C's or D's proposition would be acceptable.

In a spirit of cooperation to secure the contract, bidder A would probably be willing to split the temporary light cost as an evidence of his desire to do the work.

DECISION: Use $77,500 with $2,500 discount.

Heating, Ventilating, and Air Conditioning

(a) OUTLINE SPECIFICATIONS: Provide completely air-conditioned building, designed to recirculate air. Provide ductwork and window enclosures; concrete equipment bases. Quote lump-sum Alternate #1 price for temporary heat and maintenance.

(b) BIDS

		Remarks
A	$ 80,000	(Error?)
		Includes Alternate #1
B	$103,800	
	+300 (openings)	Contractor's estimates
	+900 (hoisting)	
	$105,000	
	+4,000 (Alternate #1)	
	$109,000	
C	$105,000	
	+4,500 (Alternate #1)	
	$109,500	
D	$101,500	
	+1,000 (equipment bases)	Contractor's estimate
	+3,500 (Alternate #1)	
	$106,000	
E	$ 78,000*	Excludes sheet metal
F	$ 23,500*	Sheet metal only
	+3,500 (Alternate #1)*	
	$105,000*	

(c) ANALYSIS: In arriving at the figure to use for the heating, ventilating, and air-conditioning estimate, we have assumed that bidder A was contacted and confirmed the correctness of his $80,000 bid. The contractor has had enough experience with bidders B, C, and D to believe that their estimates are reasonable and trustworthy, and he does not want to gamble that newcomer A's bid may be correct. Actually, he believes A to be quite wrong in that his figure does not represent a fair target toward which a formal award could be directed.

If the contractor is called to discuss his proposal with the owner, he is in a position to point out that his bid would be $25,000 lower if A's bid had been used and that some competitors may have used this figure. The owner at this point usually would recheck the estimate of the other general construction bidders. There is always a chance that the owner would be willing to accept the gamble of using A as the heating, ventilating, and air-conditioning subcontractor and would reduce the contractor's proposal accordingly. In such a case, it should be done only after the contractor has made clear his reasoning in not using A's estimate and the owner has accepted responsibility for A's performance.

The prices have all been adjusted to include Alternate #1 for temporary heat and maintenance by steam fitters and sheet metal workers. A specification for this work describing its extent and details of piping and/or the partial use of permanent piping can pay dividends in operating the job.

This might not be included in the architect's specifications, other than the reference that the contractor would be responsible for furnishing proper heating during construction. As an alternative, the work can be bid competitively, as can hourly rates for maintenance, based on some predetermined number of days required in the temporary heating season. The subcontractor is in the best position to obtain concessions from the local union with less total expense than if time tickets are signed for this temporary work.

Bidder B made no provision for providing an opening to lower equipment into the boiler room, the installation of which will involve claims by the steel erector and concrete subcontractor for their work in filling in the opening.

Bidder D omitted the concrete equipment bases, and an amount should be allowed to make his bid comparable. It is quite possible that if the work goes ahead, the concrete subcontractor could install this work without increasing the final price agreed upon for the concrete work, in which case some saving could be expected from the heating, ventilating, and air-conditioning subcontractor.

Trade practices vary in different sections of the country with respect to awarding separate subcontracts for sheet metal and other major mechanical sub-subcontracts or combining in an all-inclusive mechanical subcontract.   Both methods have worked out practically insofar as physically completing the installations is concerned.

The all-inclusive mechanical subcontractor may offer more coordinated mechanical supervision (at a price), which the contractor may continue to buy until such time as his staff has experience in supervising the additional subcontractors and the working relationships have been tried out on a smaller scale.

In our case there is a slight advantage in using the E and F combination; with D's bid very close, we could expect that a $4,500 discount could be secured in the buy out.   It must be recognized that only E has quoted on Heating separately and that B, C, and D could decide that they would only be interested in performing the entire contract.

DECISION: Use $105,000 with $4,500 discount.

Plumbing and Sprinklers

(a) OUTLINE SPECIFICATIONS: Install complete soil, vent, and water line system; roof drainage; house pump: Quote lump-sum price as alternate for temporary water.   Install sprinkler service; sprinkler heads and piping for 50 percent of basement area.

(b) BIDS

	Plumbing	*Sprinkler*	*Combined*	*Remarks*
A	$28,000	$2,750		
	+1,250 (Alternate #1)	+750		Sprinkler service; con-
	$29,250	$3,500	$32,750	tractor's estimate
B	$23,750	$3,500		
	+750 (Alternate #1)	+500		Paint concealed pipe;
	$24,500	$4,000	$28,500	estimate
C	$27,500	$5,000		
	+1,000 (Alternate #1)			
	$28,500		$33,500	
D	$22,000	$2,000		
	+1,500 (Alternate #1)			
	+1,000 (excavation for trenches)		$26,500*	Excavation for
	$24,500			trenches; contrac-
				tor's estimate

(c) ANALYSIS: For comparative purposes, adjustments have been made to all bids by including Alternate #1 for temporary water as well as adjusting other bids to include the sprinkler service, painting of concealed-sprinkler piping, and excavation for trenches, all of which was specified for inclusion.   The bids of B and D on plumbing are identical and would offer possibility of arriving at a better combination price in conjunction with D's bid of $2,000 on sprinkler work. Conceivably B is more interested in the plumbing contract and would not object if D handled the sprinkler work separately.

DECISION: Use $26,500 with $1,500 discount.

Site Work

(a) OUTLINE SPECIFICATIONS: Install 6-inch concrete sidewalks as shown. Install 6-inch crushed-rock base and blacktop paving in parking lot.

(b) BIDS

	Paving	All other	Combined	Remarks
A	$5,500	$2,500	$ 8,000*	
B	$9,000	$3,000	$12,000	
C	$8,500	$2,000	$10,500	
D	$5,000 (excavator)			

(c) ANALYSIS: Contracts for site work normally performed at the end of the job would usually be deferred until more pressing trades had been handled. The excavator who also performs paving work submitted a low bid, which may not be taken advantage of since it is unlikely that he would receive the excavation contract. The bid should be kept in mind and reconsidered at a later date.

DECISION: Use $8,000.

Factors Influencing Estimate Variations

Teamwork in putting an estimate together is essential, with all individuals cooperating in furnishing information of which they have particular knowledge. This is the time when all minds need to concentrate on the particular trades involved to arrive at a reasoned and correct conclusion. A reading of the bids can familiarize the parties and frequently develops questions which a single individual might not notice.

Having reviewed the bids received and having assigned a figure for each trade on the summary sheet (Fig. 5), we find that a recheck is in order to determine whether each figure represents a reasonable proportion of the whole. For instance, the structural-steel figure of $62,000 represents $8\frac{1}{4}$ percent of the total estimate cost (which seems low). Items such as roofing, when compared with the quantities involved, should fit into a unit price category of, say, $70 per square for this trade.[4]

The types of buildings, architectural embellishments, extent of special mechanical or electrical installations, restrictive labor practices in the area, changes in the cost index, and labor productivity (seldom reflected in the tables) will all greatly affect cost comparisons with other jobs. The difference in building layouts, story heights, and the proportion

A. B. C. CONSTRUCTION COMPANY

MADE BY D. C.

CHECKED BY H. D.

ESTIMATE FOR A. B. JONES COMPANY

LOCATION

Fremont, Ohio

PAGE 1 NO. 306

DATE 8/14/74

	ITEM	QUANTITY	UNIT	UNIT COST	TOTAL	DISCOUNT	LABOR INCLUDED IN TOTAL
1	Demolition				4,300		Contingency
2	Excavation				20,000		+5,000
3	Structural steel				62,000	2,000	
4	Concrete				88,000	1,000	
5	Masonry				68,000	3,000	
6	Stonework				6,600	200	
7	Windows/storefronts				33,500	1,000	
8	Glazing				10,000		
9	Miscellaneous iron				14,600	600	
10	Ornamental metal				15,500	500	
11	Waterproofing, dampproofing, caulking				7,700	200	
12	Hollow metal				6,000		
13	Finish hardware				2,600		
14	Roofing and sheet metal				5,500	200	
15	Tile and terrazzo				17,500	1,000	
16	Lath, plaster, acoustic				32,500	2,500	
17	Carpentry, millwork				6,700	200	
18	Painting				11,000	1,000	
19	Resilient floors				3,300	200	
20	Elevator				39,300		
21	Electrical				77,500	2,500	
22	Heating, ventilating, and air conditioning				105,000	4,500	
23	Plumbing, sprinklers				26,500	1,500	
24	Site work				8,000		
25					672,100	22,100	
26		Less discounts			-22,100		
27					650,000		
28		Add contingency			+ 5,000		
		Total subcontracts			655,000		

FIG. 5   Estimate Form. (a) Subcontract Summary.

allocated to more or less expensive building functions will also substantially contribute to the wide variations.

Items such as air conditioning add to the cost in the trade category; also the design of the air conditioning can create a marked spread because of the proportion of outside air introduced compared to recirculated air. This could account for a 5 percent variation in the total building cost.

	ITEM	QUANTITY	UNIT	UNIT COST	Labor/Material		LABOR INCLUDED IN TOTAL
1	Labor and materials						
2	Tools and supplies				1,500		
3	General expense				1,500		
4	Cleaning and trucking				6,500		5,500
5	Job office and field staff scheduling—partial				16,000		16,000
6	Watching—partial				2,000		2,000
7	Plant				1,500		1,000
8	Plant rental—other				500		
9	Plant installation, repairs, transportation				1,500		1,500
10	Hoisting (net)				9,000		5,000
11	Temporary power				1,200		
12	Temporary fuel				1,500		
13	Temporary water				300		
14	Winter weather				1,500		500
15	Protection and safety				2,500		2,000
16	Sales and use taxes				500		
17	Workmen's Compensation and Social Security taxes				8,000		
18	Retirement benefits				1,500		
19	Building permits				1,000		
20					58,000		35,000
21	Subcontracts				655,000		
22					713,000		
23	Insurance				2,000		
24					715,000		
25	Profit, 5%				35,750		
26	Bid				750,000		
27							
28							

A. B. C. CONSTRUCTION COMPANY

MADE BY D. C.

CHECKED BY H. D. ESTIMATE FOR A. B. JONES COMPANY

PAGE 2 NO. 306

DATE 8/14/74

LOCATION Fremont, Ohio

FIG. 5 Estimate Form. (b) Estimate Summary.

It should be noted that even persons in the building field sometimes lose track of the additional costs, such as site work, landscaping, and paving, which can greatly increase the cost of a project but frequently (unintentionally) may not be averaged into the square foot allowances. Higher square foot building costs in some cases may be the result of including such site work.

All price information, wherever obtainable, will be helpful when the bids appear to be completely off the beam—either too low or too high—and the contractor is faced with having to verify quickly how much dependence should be placed on them.

The data in handbooks, such as the annual editions of *Building Construction Cost Data*,[5] tend to indicate averages, a wide departure from which would warrant midnight hours to determine the factors causing the discrepancies, unless the building can obviously be bracketed in some special category. All price information must be used with caution and cannot be accepted blindly. It does, however, represent one tool for the purchasing man in making his comparisons.

OVERHEAD OR BACK-PAGE ITEMS

During the estimate preparation in determining quantities and subcontract bidding, a very vital phase of the study has been proceeding, involving the pricing of overhead costs and the special factors affecting job operations (Fig. 5b). Here the prospective superintendent to be assigned to the job can be of very practical help.

Each special item should be analyzed for the specific job and priced accordingly. Only after the contractor has developed trustworthy cost records can he assign his own realistic square foot costs for items such as general cleaning, temporary light, window cleaning, carpentry costs for temporary protection, and similar general charges.

In setting up the estimate, consideration should be given to showing the special items, peculiar to this job, as part of the basic estimate. The listing of overhead items should be restricted to charges which apply consistently to every job. This will help to establish a more realistic future base for overhead and also facilitate the processing of the contractor's change estimates. The owner's representatives sometimes are reluctant to approve estimates reflecting a substantial percentage for overhead costs which the contractor knows from analysis are actually applicable to the project.

The Job Overhead Summary #8800 was prepared by The Associated General Contractors of America and provides a checklist which enumerates cost items which may or may not be apparent from a reading of the specifications.[6] The lists are extensive. The main reason for their use is to verify that provision for different items has been made in the basic estimate and that they are not duplicated in the overhead percentages.

As we have noted earlier, too much reliance cannot be placed on square foot or cubic foot building costs. It must be recognized that, although the practice is often followed in the early development stages of a project, closer estimating is always needed when lump-sum bids are being prepared.

For the record, the American Institute of Architects (AIA)[7] has published guidelines noted as Document D101 to define area and volume with respect to buildings, as follows:

ARCHITECTURAL AREA OF BUILDINGS

The ARCHITECTURAL AREA of a building is the sum of the areas of the several floors of the building, including basements, mezzanine and intermediate floored tiers and penthouses of headroom height, measured from the exterior faces of exterior walls or from the centerline of walls separating buildings.

Covered walkways, open roofed-over areas that are paved, porches and similar spaces shall have the architectural area multiplied by an area factor of 0.50.

The architectural area does not include such features as pipe trenches, exterior terraces or steps, chimneys, roof overhangs, etc.

ARCHITECTURAL VOLUME OF BUILDINGS

The ARCHITECTURAL VOLUME (cube or cubage) of a building is the sum of the products of the areas defined above (using the area of a single story for multistory portions having the same area on each floor) and the height from the underside of the lowest floor construction system to the average height of the surface of the finished roof above the various parts of the building.

The user should ascertain the latest edition when using AIA forms and documents.

It should be noted that these definitions do not reflect varying percentages of usable space nor in themselves reflect the amount of interior partitioning and finishing which might well create a 25 percent variation in the total project cost.

Much price information is published periodically in trade magazines such as *Engineering News-Record, Plant Engineering,* F. W. Dodge *Reports,* and *Contractor News,* which help to spot regional variations. Some contractors carry this one step further and check with their major suppliers, subcontractors, and manufacturers, bimonthly or quarterly, to determine the current status and future trends of pricing in their particular locality. The result can be and usually is a greater price consciousness regarding building products commonly used.

All contractors submitting proposals tend to regulate their own estimating by the time available, usually planning to meet the day before submission to analyze subcontract bids and clear up unanswered questions by the bidders which might change the amounts quoted. (Naturally, this schedule varies with the size of the project.)

The sooner the subcontract bids are received, the greater the opportunity to analyze and discuss methods and possible savings with the bidders. This points up the desirability of someone in the contractor's office keeping in close touch with all bidders to recognize their problems and to keep an eye on their progress. Courtesy subcontract bids or "no shows" are no help to the contractor striving for a contract.

ESTABLISHING THE FEE

The fee or profit on a job will have been considered in the early stages of the estimate preparation, based on the job's scope and duration and the availability of the contractor's organization. Whether it needs to be 10 percent of the estimated cost or as high as 25 percent because of its small size, greater requirements for supervision, or alteration problems, the contractor will decide from his past experiences in the marketplace and will determine the fee to be added by his interest in this particular job.

If the building for which the contractors have rechecked their estimates is more or less typical, the likelihood is that the costs will be reasonably close since they have been prepared using verifiable quantities and probably a similar range of subcontract bids.

The fee or profit on the job will vary, and with it the final amount of the proposal, depending on the contractor's interest in submitting a tight bid and his analysis of the bidding procedures of his competitors (Fig. 5*b*).

Being a low bidder consistently is not desirable. Over the years, the low bidders are the first to go bankrupt. Each job should be bid to show a profit. There is little merit in securing contracts for the sake of volume, whatever the reasons may be for proceeding on this basis.

RECHECKING THE ESTIMATE

Once the estimate has been put together and totaled, a recheck is an excellent idea. Even though bids have been reconciled and appear to be in order, some trades may appear to be consistently above average based on percentage comparisons of total building costs which might have been expected.

Frequently these variations are due to less efficient use of space and can only be corrected by layout or design changes after further study by the architect and owner. To the extent that the contractor can spot possible revisions for economy and propose alternatives, he can help significantly to bring a job within its originally anticipated cost.

In a Third Quarterly Cost Roundup featured in *Engineering News-Record* of September 21, 1967, an article appeared entitled "Sound Area Analysis Is Crucial to Controlling Costs of Buildings."[8]

The material in the article, with examples, was developed from studies made by Wood and Tower, Inc., construction-planning and cost-control specialists of Princeton, New Jersey, and reflected experience on several types of buildings located throughout the country.

This firm, which also supplies data for cost monitoring in the *Dodge Manual for Building Construction, Pricing and Scheduling,* has updated and expanded portions of the *Engineering News-Record* article to include the following, which is published with the permission of Wood and Tower, Inc.:

Average Ratio Chart

Structure	Average ratio (%)*
Warehouse	111
Department store	127
Library	136
Bank	145
Administration	154
Classroom	157
Dormitory	159
Laboratory	176
Hospital	189

* Average ratio = gross sq ft/net sq ft.
SOURCE: Wood and Tower, Inc.

To facilitate understanding and comparison of estimates, the following format has been developed by Wood and Tower, Inc., for building "systems":

Construction costs	*Added project costs*
Foundations	Fees
Floors on grade	Surveys
Superstructure	Borings
Roofing	Inspection
Exterior walls	Testing
Partitions	Legal expense
Wall finishes	Administrative expense
Floor finishes	Movable equipment
Ceiling finishes	Interest on investment
Vertical transportation	Land-acquisition costs
Specialties	Share of central utilities
Heating, ventilating, and air conditioning	Design development contingency
Plumbing	Project contingency
Electrical	
Site improvements	
Fixed equipment	
General conditions	

Finally, regional differences in construction costs, essentially a reflection of wage differentials, have been plotted for some of the selected systems in terms of a United States base.

14 Regional Cost Differentials

Region	Architectural and structural	Heating, ventilating, and air conditioning	Plumbing	Electrical
Atlanta, Georgia	.93	.97	.96	1.01
Birmingham, Alabama	.83	.93	.95	.91
Boston, Massachusetts	1.05	1.02	1.05	1.06
Chicago, Illinois	1.03	.99	1.02	1.06
Dallas, Texas	.91	.95	.91	.94
Denver, Colorado	.95	.97	.93	.95
Detroit, Michigan	1.05	1.04	1.06	1.11
Los Angeles, California	1.03	1.04	1.06	1.01
Miami, Florida	.98	1.03	1.02	1.02
Milwaukee, Wisconsin	1.00	.99	1.01	1.03
New York, New York	1.16	1.08	1.14	.98
Omaha, Nebraska	.93	.98	1.00	.98
Portland, Oregon	.98	.97	.99	.93
St. Louis, Missouri	1.00	1.02	.98	1.03

SOURCE: Wood and Tower, Inc.

Invariably, when analyzing bids and preparing an estimate, there will be occasion to telephone some bidders to request clarifications. The contractor should not disclose unique or innovative methods developed by one bidder to others competing. Fair treatment of this type is appreciated and can be counted on to pay off in future negotiations by the purchasing man.

Throughout the preparation of an estimate, it is well to maintain a sense of proportion—of the importance of new information received in relation to the overall project. The change in size of the medicine cabinets required for a new hotel would not significantly affect costs, whereas the fact that the ceramic tile walls will be full height rather than 5-foot-wainscot height could increase costs significantly as well as affect several trades.

As the jobs become larger, more individuals, both within the contractor's office and outside, become involved in transmitting information that should be considered confidential. It should be recognized that occasionally there are "leaks," and safeguards need to be agreed upon in advance to minimize the risks of disclosure.

For example, when estimating insurance costs, the broker should be asked to quote rates rather than submit a total estimated premium for a specific contract price. When presenting proposals for public work, a corrected proposal, submitted exactly when due, may adjust figures which were susceptible of previous knowledge by others.

Before the proposal is finalized, regardless of the size of the contractor's organization or the particular job being estimated, it should be reviewed by the contractor personally or by the executive in charge—someone not closely involved in its preparation—to satisfy himself that it is correct, that the quantities for one story have not been omitted, or that one sheet of the estimate has not been omitted from the summary.

It is hard to forget the case of a Midwestern contractor some years ago, who had bids of $625,000 each for plumbing and for heating from a subcontractor and used a total figure of $625,000 for both in his estimate, based upon which he was awarded the general contract.[9]

FINAL SUBMISSION OF THE ESTIMATE

The period immediately before the submission of a lump-sum proposal on a large job quite often is hectic, with respect both to bid analysis and unintentional or intentional delayed submission of subcontractors' prices, and to actual preparation of the estimate, execution, and final delivery to the office receiving the proposal.

It is of the utmost importance, especially for public openings, that all proposal deliveries be on time. On occasion, even a few minutes' delay may cause rejection of a proposal. While some discretion may be granted by the persons authorized to receive proposals, it cannot be assumed. It can be expensive as well as disheartening to fail to have a bid considered after much effort in its preparation just because it was presented a few minutes late or because of some minor irregularity in form.

After a job has been awarded to another contractor, information obtained on public openings (frequently published in trade papers) or on private work, through discussion with the owner, is always of value to those involved in the preparation of future estimates. It establishes the state of the market and prompts the contractor to consider future changes he might make in his own estimating to approximate the lower bids. In addition it can help to keep channels of communication open with owners so that a mutual understanding and respect may be developed for the contractor in the future.

With all due respect to the summaries of bids frequently listed for much public work, there are dangers in assuming that building contractors will always submit bids consistently in the future. Errors may occur, contractors may find themselves "overloaded," or different bidders may enter the picture.   Bidding contingencies should be analyzed care-fully, but beyond that there are real dangers in basing estimates on mathematical "probabilities" without a full recognition of the changing labor and material market, the state of the drawings and specifications, and the job in question.

During the preparation of *Builder's Guide to Contracting* there have been substantial increases in subcontract costs.   While all building cost indices fluctuate, the components may not change uniformly.   Variations from very sharp to nominal changes make it essential when projecting costs to study the component trades rather than arbitrarily assume any across-the-board trend.

NOTES—Chapter 2

1. Appeal of F. H. Antrim Construction Co., Inc., AG BCA #307, 72-2 BCA 9475 (1972).
2. George F. Sowers, "Changed Soil and Rock Conditions in Construction," *Journal of Construction Division: Proceedings of the ASCE*, November 1971.
3. No reflection on metal industry bidders is intended.   The example cited applied to another trade.
4. Robert Snow Means Company, Inc., "Building Construction Cost Data," Duxbury, Mass., 1973, pp. 100, 246.
5. Dodge Building Cost Services, McGraw-Hill Information Systems, New York.
6. Job Overhead Summary #8800, The Associated General Contractors of America, Washington, D.C.
7. This definition has been reproduced with the permission of The American Institute of Architects. Further reproduction is not authorized.
8. "Sound Area Analysis Is Crucial to Controlling Costs of Buildings," *Engineering News-Record*, Sept. 21, 1967.
9. Guy C. Kiddoo, "Loans to Contractors," *Bulletin of Robert Morris Associates,* April 1952.

CHAPTER THREE

Receiving Bids, Budget Preparation, and Proposal for Negotiated Work

As every general contractor knows, work is not simply awarded on a silver platter. Jobs must be estimated competitively, won, and sweated over until completion. There is a never-ending succession of problems as the contractor cuts his eyeteeth through experience and hard work.

From the owner's standpoint, the lump-sum general contract usually represents the best way to perform construction unless there are controlling factors against its use.

A definite agreement covering the price and terms for a specific scope of work can be agreed upon. This presupposes, of course, that the drawings and specifications have been completed and that there is little likelihood of major changes.

Unfortunately this is not always the case. Production plans in the factory may change; companies may merge; the tenant that was to occupy certain space will no longer be leasing, and the new tenant will enter only if major changes are made to accommodate him. New buildings, partially completed, suddenly require a full basement, creating an altogether new ball game as far as safeguards which an owner might expect with a straight lump-sum contract. These things happen, and if there is any indication that the plans are not jelled but that the necessity to proceed still applies, it is far better to arrive at some other form of general contract.

In this chapter, we will devote time to analyzing the types of jobs that lend themselves to negotiation rather than lump-sum competition. As a general rule, a negotiated contract is appropriate whenever the advantages of proceeding with the work at once outweigh the fact that (1) the drawings and specifications on which to base a firm price are not sufficiently complete and (2) the manner in which the work would be performed is subject to considerable change in order to accommodate existing and developing situations.

It is particularly desirable that the contractor endeavor to obtain some negotiated work and steer clear of acquiring a reputation which W. R. Park classifies as a "price cutter" or "bidding fool."[1,*] If price only will be the criterion, lump-sum bids should be prepared only when there is a reasonable chance of securing the award. If the notorious low bidder of the area is involved, it would be well to avoid the job and concentrate on estimating some other project.

Conditions vary widely, but six possibilities can be listed. These include cases where there is need:

1. To continue in the present location while maintaining operations during construction. This could apply to retail establishments and small manufacturing plants.

2. To expand current production facilities quickly. Time is of the essence, and architectural planning may have to proceed concurrently with initial construction work, with priority given to incoming and outgoing shipments of products.

3. To be alert and prepared to furnish services promptly to repair damages due to fires or floods, at the same time incorporating new ideas that had not been thoroughly crystallized because of the unexpected emergency that has arisen. If some individual in the contractor's organization has the professional engineering qualifications, he can be involved in the preparation of appraisals for insurance purposes. There have been occasions when this work has become a major source of revenue for concerns, providing steady, year-round income.

4. To start construction immediately to secure greater flexibility in the development of the office, plant, or store because of contemplated zoning changes which might impede the new work.

5. To coordinate manufacturing operations with construction to a considerable degree, with obvious advantage in performing some construction and subcontract work on a cost-plus basis.

6. To furnish budget estimates and continuing-cost studies to the architect and owner concurrently to permit decisions with respect to types

* Superior numbers refer to Notes at end of chapter.

of materials, labor availability, recommended areas for savings, and construction methods. Time schedules, estimated lead times, manpower requirements, and cash flow can be helpful.

Normally chain-store management would prefer to engage competent local contractors. If these are not available in the area or if, in receiving proposals the amounts appear quite unreasonable, the contractor who knows the ropes occasionally will be in a better position to quote advantageously, even though he is away from his home base of operations.

In many cases, each succeeding store becomes easier to price accurately because of the accumulation of reliable cost records resulting from the similarity of the basic construction in the chain-store system.

Some projects such as the modernization of railroad, bus, or air terminals also involve considerable interference with large surges of pedestrian or passenger traffic which cannot be completely rerouted and must be contended with. Usually only 4 or 5 hours a day can be devoted to the critical work area, and the balance of each day must be devoted to performing some kind of productive work in other areas.

All these circumstances lend themselves to the development of a working arrangement in which the contractor is teamed up with the architect to fulfill the owner's requirements. In this type of relationship both the architect and the contractor need to have the requisite skills to operate with care, using their best judgment to serve the interests of their client, the owner, honestly. For the contractor, a negotiated contract becomes appropriate and equitable.

BUDGET ESTIMATING

Budget estimating differs from that covered in Chapters 1 and 2 in several respects. Let's spell out the differences.

1. Usually only a brief outline specification is available for the estimate. The budget should be clearly referenced to the data actually used in the bidding. Items which the contractor recognizes from his experience as being required in addition, if any, should be included, with reference made to each and its costs indicated.

2. Where a portion of the work could be subcontracted on a lump-sum basis, budgets should be obtained from one or two qualified bidders. Presumably this work could be done during normal working hours, without major interference with other trades or the owner's operations. It might involve new masonry partitions, toilet room work, flooring, etc.

3. Where subcontract work might lend itself to performance most economically on a cost-plus basis, because of inability to clearly define the scope of the operating conditions, the budget estimate should include a sufficient contingency to cover interferences, overtime expense, and incidentals to complete the work even though these may not be com-

pletely outlined. There is no advantage to owner or contractor in competitive bidding on budget estimates. The need is to secure a trustworthy appraisal rather than to arrive at an obviously low or incomplete figure.

4. Usually a qualified mechanical and an experienced electrical subcontractor will be asked to budget their work. Selection of these bidders should be made with the concurrence of the owner and architect. Such bidders should have the necessary experience in budgeting even though some provisions may not have been completely spelled out.

5. The timing of the work is important. For instance, the estimate might be predicated on field work starting one month after the receipt of the budget. Should the decision on the award be postponed for a longer period, estimated escalation cost during the period would represent a justifiable increase in the budget. The owner, of course, is interested not only in "How much will it cost?" but also in "How much will it cost if I wait?" The projected approximate completion date should be available to the budget bidders to appraise the wage rates during the life of the project realistically.

Later, if more-detailed specifications become available, a quick comparison should be made with the outline to determine if more-expensive materials or more-restrictive requirements have been proposed than those which had been budgeted. Not infrequently, mechanical and electrical subcontractors, and general contractors as well, budget work only to find that their assumptions have now been "gold-plated." These variations need to be pointed out at once to owner and architect. Possibly the embellishments can be waived in the interest of keeping within the owner's proposed budget. Naturally such items would increase the amount of the guaranteed cost of the job if the contract is so prepared. If the owner is forewarned, there may be a decision to forgo the change before it is incorporated in the revised drawings.

If the client's ideas have not developed to the point of plans for bidding, some owners prefer to engage a contractor and reimburse him for estimating the probable costs. In such cases, the owner usually makes it very clear when sending out the project for bids that the work will not necessarily be awarded to the contractor who prepared the preliminary estimate.

Frequently the owner will simply select a contractor, outline his requirements and the status of the sketches and/or specifications, and ask for an estimate in 4 to 6 weeks, without offering to cover the contractor's costs. There is always a serious question of who is doing whom a favor when the contractor estimates such a job. A mutual spirit of helpfulness can produce real benefits on both sides of the table. The owner should realize that estimating costs represent an important part of the contractor's yearly overhead. When budgeting services are required, many owners will appreciate what is involved and will make an effort to reciprocate in some way.

Whether his time is reimbursable or not, the contractor selecting subcontractors to perform budgeting work not only should be acquainted with their capabilities and able to discuss their qualifications with the owner, but also should know of their availability for the proposed job.

A QUESTIONNAIRE FOR SUBCONTRACTORS

Several kinds of information should be at hand when major subcontracts are involved and the contractor is considering or suggesting the possibility of cost-plus work.

1. Current subcontracts with percentage of completion. (Would the subcontractor be overloaded if he were awarded the proposed work?)

2. Subcontracts recently completed; experience in this locality with references.

3. Volume of work which the subcontractor has previously performed on a cost-plus basis.

4. Extent of work which would be sublet: sheet metal, pipe covering, thermostatic controls, rigging, etc.

5. A recent financial statement noting past-due accounts, taxes withheld and owing, and working capital.

6. Shop facilities available for fabricating special requirements for this particular job.

7. Special equipment available for field operations.

8. Supervisory personnel available for this work and their previous experience on similar projects.

In the case of the mechanical and electrical trades, where interferences and relocations are likely to complicate the work, an experienced subcontractor preparing budget estimates is sure to include an allowance for delays and similar problems. While in such cases a cost-plus arrangement offers a practical solution, when the budget estimate is prepared, there is no advantage in deciding immediately that the work must be paid for on this basis. Rather, these trades should be expected to budget their work realistically, arriving at an estimate that would include contingencies for special circumstances. Later, in Chapter 5, details of these subcontracts that could be negotiated will be reviewed.

It is especially important both from the architect's and contractor's standpoint that the owner is not enticed into decisions to build because of incomplete or misleading information. Aside from the irreparable damage to one's own reputation, the owner's financing could be seriously impaired, with consequent risk to other parties. A mutual trust, a recognition of the types of situations coming up for decisions by an owner's building committee that require study before irrevocable commitments are made, and an effort to serve the owner most conscientiously are essential for a healthy working relationship that could lead to repeat business.

From the contractor's point of view, it is important that the budget estimates he furnishes be accurate and that the provisions—namely, what is and what is not included—be spelled out in detail so that there is no possibility of a misunderstanding on the part of the owner or architect. The contractor's reputation can be made or broken by the attention, care, and skill used in the preparation of these cost studies.

Understandably, budget estimates are not always acted upon immediately. If the development time of the job is extended, it is important that the estimate be checked periodically to flag such things as deteriorating market conditions or experience on other projects which could call for a revised judgment in the pricing of the previous study.

Another consideration that bears on budgeting costs for this type of job and that affects the overall success of the project should be noted. New production methods and special equipment—sometimes very expensive—may have been selected and estimated by the owner's production staff. The field may be unfamiliar to the contractor, and the tendency would be to accept the owner's evaluation without question. The contractor who digs deeper, especially with an inexperienced owner, will try to recheck the analysis and lend a helping hand in pricing. There is little satisfaction in having the entire project overrun because of some poor estimating, even though the contractor had no part in its preparation.

OWNER'S BUILDING DEPARTMENTS

There are times when an owner's building requirements are so extensive and involved that it may be practical to set up an operating department to handle construction, or even, less frequently, to organize a separate construction company to serve as a general contractor.

It is not uncommon for owners to subcontract major alteration or remodeling projects directly through their building or purchasing departments and perform the function of a coordinating general contractor.

Where there is likelihood of this method being followed, many of the procedures used by the contractor would be equally applicable to an owner's subcontracting operations in similar situations.

TYPICAL ALTERATION JOB

For our illustration (a job which might well be handled under a cost-plus general contract), we will consider a one-story-and-basement factory building, 50 × 100 feet, for which there are compelling reasons to expand its area by 100 percent rather than to relocate. It has been decided that a second story should be added over the existing building and that a 50 × 100 foot one-story section should also be added. Two rows of columns along the length of the existing building will have to be replaced by a single row to accommodate new heavy-duty manufacturing equipment. The basement will now be used for light manufacturing, while the second floor will accommodate office and engineering departments.

With the alteration work involving excavation and new foundations in the existing building, there is bound to be some interference with the owner's operating departments. To lessen the impact, significant amounts of overtime on the part of the construction crew is a strong possibility. When the contractor has men to perform this work, it is usually more efficient and less expensive to do so than would be the case if it were performed on a lump-sum-subcontract basis. For this reason, we are assuming estimates on the basis that demolition, excavation, and concrete would be handled by the contractor.

For illustrative purposes, we assume that the work of some of the subcontract trades is clear-cut, that interference with other trades is minimal, and that a lump-sum-contract price is quite possible. In such cases, the budget estimates can be gauged realistically and noted on the estimate sheet. If and when the job is released for construction, an effort should still be made to secure at least two other firm competitive bids in each trade before making an award.

The case of the erection of structural steel is complicated by the need for cutting through floors and roofs. If this cutting is performed by the general contractor, then the erection work becomes more standard, and lump-sum contracting could be expected in this trade. This presupposes that a thorough study of the preparatory work has been made by the contractor and his costs determined for this and other conditions involving "access" to perform the work in a practical sequence of operations.

An alternative method, if there is a lack of competition for the work, would be to secure lump-sum prices for the delivery of fabricated steel to the site and to have the fabricator also perform the erection work, but on a cost-plus basis with a total cost not to exceed some guaranteed sum. This would avoid the situation where the erector includes a substantial contingency in his estimate for a lump-sum contract to cover expensive working conditions which might not occur.

The second-floor steel-framing system would be erected over the roof of the existing building, with existing-roof demolition to follow later.

The excavation and concrete have been estimated in detail and quota-

FIG. 6 **Invitation to Bid (letter form).**

X. Y. Z. Construction Company
North Street
Knoxville, Tenn.

September 5, 1974

Re: Proposed addition to
Plant Building
Rocky Mount, N.C.

Arch./Engr.—Smith & Smith
Mech. Engr.—Brown & Jones
Charlotte, N.C.

To All Bidders:

You are invited to submit your proposal for work in your trade in connection with the proposed addition to the plant of Randall Brothers, Rocky Mount, N.C., in accordance with drawings and outline specifications both dated September 2, 1974. For your convenience, data are available for estimating purposes at the offices of Smith & Smith, architects, Jones Street, Charlotte, N.C.

Your proposal, in duplicate, is requested at this office by noon, October 2, 1974.

Bidders should note: All cutting of walls, floors, and ceilings within the existing building will be performed by the general contractor except where such cutting is claimed by a specific subcontract trade, in which case the cutting shall be included in the subcontract bids. All patching of finishes in the existing building will be estimated by the general contractor.

Breakthroughs for the new addition to the existing building will be made by the general contractor. The various subcontractors are to include the cost of patching dividing walls to match existing finishes.

Heating, ventilating, and air-conditioning subcontractor and electrical subcontractor are to reroute mechanical piping, ductwork, and conduit as required and shall indicate separate allowance included in the proposal for such rerouting.

Proposals shall be based on performance of the work during normal working hours for the trade.

Our standard subcontract form which includes a broad form hold harmless clause will be used for the work.

All questions which you have or requests for the use of drawings should be directed to the undersigned, at our Knoxville office (Telephone # _____), who will be at the plant site on Monday, September 10, 1974.

J. J. Smith
Purchasing Agent

tions received from vendors and material suppliers. We do not preclude the possibility of making a subcontract package for some of these trades, but at this stage it seems more likely that the contractor will handle it directly.

We will assume that the mechanical and electrical work are complicated. The specifications, while listing the general scope of the work and the approved manufacturing sources, neither give sufficient detail for close estimates nor spell out interferences fully. "As built" drawings are not available, and we assume that much of the work cannot be inspected.

Assuming that the necessity for cost-plus work has been accepted by the owner and that there is no objection to using contractor A for mechanical work and contractor B for the electrical trade, the general outline of the work should be explained to both. The possibility of cost-plus work should be discussed with the further understanding that, if the work is authorized, A and B would be the subcontractors. On this basis, it could be expected that both would lend their best efforts to the preparation of the budget. The mechanical engineer meanwhile has consulted his records of previously designed jobs and should be in a position to approximate costs when the results are being analyzed.

In selecting the subcontractors to do this estimating, only those should be utilized who are willing to devote the time and who will continue to be available for subsequent cost comparisons which may be required in the initial period. A replacement bidder may not necessarily see the job in the same perspective.

Once the decision has been made to proceed on a cost-plus basis, at some stage and before the subcontractor's figure is finalized for presentation to the owner, a copy of the contractor's typical cost-plus agreement, as well as a list of the general operating conditions for the job, should be available to the subcontractor for examination. The amount of the fee and any profit-sharing arrangements may be subject to later negotiation, but the separation of the allowable reimbursable costs and the items to be included in the fee should be recognized in the preliminary negotiating phases. Charges for rental of tools and the use of the subcontractor's small equipment, engineering coordination charges, purchasing charges, and similar overhead expenses should be definitely allocated and, if possible, included in the subcontractor's fee.

PROPOSALS FOR THE WORK

Experience in preparing a budget proposal includes a recognition of the condition of the bidding information. If the data are sketchy, obviously a contingency should be added and so stated.

Even though the information may appear to be reasonably complete, the summation of the various pieces frequently will not represent a reliable total. On overhead items, for example, such as protection involving many classes of carpentry work, there may be a tendency for the final costs to be considerably more than was budgeted for the individual items. In this case, the contractor's estimate for other similar jobs—on a square foot basis—may be a more-reliable criterion.

After the estimate for the various trades is totaled, as was done with our lump-sum estimate, the proposal for the negotiated work should specifically state those items which are not included. These may be provided by either the owner or the contractor later. Frequently, owners arrange for both the purchase and installation of the production equipment as well as for procurement of special finishing items such as draperies or murals in specially finished offices.

The list of exclusions tends to be the start of a checklist to ensure that all costs are being considered.

There usually is less formality in the presentation of a proposal for work of this nature and more opportunity to discuss various aspects of the job which offer the possibility of alternative treatments.

Similarly, the subcontract work which may be handled under a cost-plus contract should be thoroughly discussed with the subcontractor to determine alternatives which may offer possible economies or revisions about which the owner should be informed in the consideration of the proposal.

The contractor needs to recognize that manufacturing plants are in business to produce finished products at a profit. Construction—especially when it must be performed simultaneously with maintaining production—is decidedly a nuisance.

If the layout is such that the architect has considered having the new facilities (toilets, stairways, and elevators) become available in a sequence to permit their use, followed by demolition of the existing areas, the contractor can—with step-by-step planning—outline construction procedures that may appeal to the owner, indicate an appreciation of the manufacturing problems that the owner is concerned with, and make the selection of cost-plus contracting more certain.

OTHER METHODS

There are other methods of handling complicated building operations but, as they lend themselves to larger operations, we will touch on them only lightly.

In some cases, especially with special operations in the chemical, oil-distilling, and paper- or steel-manufacturing fields, so-called "turn-key" contracts provide for the complete design, construction, and contract administration of the project from start to finish. While there may be circumstances where this method has advantages, for the smaller contractor whose "bread and butter" comes through working cooperatively with an architect and owner, this type of contract offers rather minimal opportunities.

CONSTRUCTION MANAGEMENT AGREEMENTS

Modern building construction during the past decade has become so large and complex that there has been a dearth of professional talent able to design, coordinate, and contract for the construction of large projects ranging from $50 million or more in a manner to conserve time and to minimize and guarantee costs for an owner—public or private.

The result has been a definite tendency toward construction management and project administration where arrangements are made with selected qualified firms to coordinate the design and construction process, including the operation of the separate contractors and subcontractors, frequently through some form of agency contract.

Under other names, the practice has been followed by the larger contractors for many years where the drawings were not sufficiently complete for lump-sum bidding but where there was a necessity to proceed with field work.

Conditions vary, but generally speaking the advantages are apt to apply and less duplication of effort is involved if the projects are in excess of $5 million.

On occasion, the contractor can serve as the construction department for local manufacturing plants. This arrangement, if developed over the years and carefully nurtured by experience and integrity, offers good possibilities for a profitable relationship without the business risks involved in performing only lump-sum contracting. Some contractors find that a steady, even though small, volume of work in such a plant can help toward general overhead expenses and does not require an unusual amount of supervision.

While some contractors prefer to restrict their work to larger contracts, many prefer to perform tenancy requirements or changes on buildings which others have constructed. This class of work, while competitive, lends itself to a close working relationship with the building owners, and it offers the smaller contractor the chance to familiarize himself with an owner's operations and learn how best to make his organization useful. In many cases such work is not subject to the uncertainties of the weather.

Some larger companies, especially in the metropolitan areas, find it profitable to handle this work under a separate department, with minimum overhead and geared to serving an owner's needs most expeditiously.

It is, of course, possible that the contractor's budget studies may have indicated to the owner that a complete new building would be justifiable, in which case the same reasons for a negotiated contract no longer apply. Instead the job may be simplified, and a lump-sum general contract may become the best agreement under which to proceed.

To the extent that the contractor has assisted the owner in reaching a decision to modernize and expand or to build, he has taken the first step in creating a business relationship that could be helpful with other prospects in the future.[2]

Of course, the first consideration is the basic management decisions regarding the production forecasts—its objectives and present and future space needs. The transportation problems, community patterns, utility rates, and labor trends will all have to be evaluated by management.

Physically, the floor loads, column spacing, clearance heights, nonparallel walls, building shape, windows, and door openings are factors that can affect the usefulness as a warehouse, light manufacturing plant, or office. The contractor is in a good position to assist in surveying these conditions for a further study of the existing facilities.

In comparing the new plant and the existing facility over its life, the space unit cost points to the best choice. To be decisive, the difference between the new and the modernized facility should be more than 10 percent.

Concurrently with the production forecasts, the management will find it necessary to analyze the needs of the company and the factors that are of importance in determining proper site selection. It's a "task force" job, with several hundred questions requiring answers before a considered decision can be arrived at.[3]

NOTES—Chapter 3

1. W. R. Park, *The Strategy of Contracting for Profit*, Prentice-Hall, Englewood Cliffs, N.J., 1966.
2. "Modernize or Build," *Factory*, October 1963.
3. "Plant Site Selection Guide," *Factory*, May 1957.

PART TWO

Negotiations with Subcontractors

CHAPTER FOUR

A Lump-Sum General Contract and Preparation of Subcontracts

Once a contract has been awarded to a contractor, the areas of responsibility in his organization need to be confirmed promptly. Our emphasis will be on the arrangements for subcontracting the work and purchasing materials and equipment. Additional functions of the contractor such as acting as liaison between the owner and architect, performing accounting work, providing insurance, checking job safety provisions, and processing plan and field changes will all have to be dovetailed with the purchasing and subcontracting decisions as the work progresses. The purchasing agent should be completely familiar with these phases since in some companies he will be expected to handle this work as well as his own procurement responsibilities.

Primarily, the contractor wants workable agreements with his subcontractors which will adequately provide for the performance of the work in accordance with the drawings and specifications, at a contract price no higher than the estimate used to compute the general contract price.

TARGETS IN NEGOTIATING

Before moving along with the individual commitments on subcontracts, the contractor should have a clear overall conception of the subcontract prices needed to meet his own estimate. Using the estimate form (Fig. 5) in Chapter 2, we assumed discounts of $22,100 for an estimate of $750,000. To average the computed percentage of approximately 3 percent, the larger subcontracts potentially offer the best possibilities to obtain the necessary savings. For instance, the chances of having a $5,000 saving on one $100,000 subcontract are likely to be better, and it would be less time-consuming to negotiate one large subcontract than a group of minor subcontracts.

PRIORITIES

Priorities have to be established. How soon can or should the excavator start? What other trades should be under contract? Are temporary services required or could they be deferred until the award of the major mechanical and electrical subcontracts? How soon should the elevator subcontract be awarded to ensure operation by the scheduled completion of the building? Are the structural-steel framing members of a size whose delivery could be adversely affected by the steel-mill rolling schedules? Both field and office should share in preparation of the list of contract items (Fig. 7), which represents a good initial form for the outline of scheduling information.

MADE BY	A.B.C. CONSTRUCTION COMPANY				CONTRACT	
DATE					CONTRACT NO.	
	LIST OF CONTRACT ITEMS					
REQ. NO.	DESCRIPTION	DELIVERY WANTED	REQUISITION		SUBCONTRACTOR/VENDOR REMARKS	
			WANTED	ISSUED		

FIG. 7   List of Contract Items.

The bar chart (Fig. 4) referred to in Chapter 1 needs to be rechecked with the actual calendar starting and completion dates to show the projected completion of the building. Later, after the major trade contracts have been awarded and the starting dates for steel erection accurately established, the chart can be refined to confirm the final progress schedule for all the trades.

When the work was first estimated, it may have been necessary to accept some subcontract bids simply on the basis that they appeared reasonably correct based on past experience and the range of the bids.

Now is the time to analyze the various estimates and confirm that the relationships are consistent with earlier contract awards.

The contractor has a knowledge of the range of the subcontract bids and also of which trades should be negotiated first for the benefit of the project. Subcontractors recognize this situation and usually give prompt attention in responding to all inquiries about their proposals. It is good practice to talk with all bidders and to avoid any preconceived ideas regarding an award until all have had a chance to discuss their proposals. Interested subcontractors will always appreciate this consideration. After this has been done and the contractor has the "feel" of the approximate best prices obtainable, he can make a better determination of what the work may be bought for.

Usually when there are several subcontractors to interview, the purchasing man would ask the bidders submitting the higher bids to come in first to discuss the job, if this is feasible. This permits a certain amount of "gleaning" of information which may or may not be of value when interviewing the lower bidders who it may be assumed are more likely to be in the running.

Where conditional awards have been negotiated prior to submission of the contractor's proposal, a meeting now should be held to conclude terms of the agreement.

Serious negotiations should not be entered into until the contractor is ready to buy. At this stage subcontractors are most interested in concluding an agreement. Assuming a subcontractor has reanalyzed his estimate, he would not be likely to offer any further major concessions thereafter; and if negotiations drag on, interest usually decreases.

Presumably the subcontractors were invited to bid because of satisfactory performance in the past. This should be verified during the bargaining period with calls made to check on recent performances. A background of similar, satisfactory contractor-subcontractor relationships can make negotiations proceed more quickly, just as good field experience and contractual arrangements will make the subcontractor more receptive to another venture.

NEGOTIATING POLICIES

All construction organizations need to have policies and guidelines agreed upon by the executives and set forth as the criteria by which the company is to operate. It is not a matter of being altruistic as much as it is a recognition that there must be standards of conduct which apply to the representatives of the contractor who deal with subcontractors. Over the years these business relationships establish a reputation that can significantly affect the efficiency of negotiations and the ability of the contractor to obtain more advantageous terms and subcontract prices.

To encourage such mutually rewarding relationships with subcontractors, it is worth keeping in mind that while every effort should be made to obtain the best terms for the contractor, any agreement should also be fair from the subcontractor's point of view. Advantage should not be taken of apparent errors in the subcontractors' bidding estimates. Individual subcontract bids should not be quoted to others.

Subcontractors recognize that it will not be possible to secure all the work being bid; they will be unlikely to receive even 25 percent of the jobs studied. At the same time, when a subcontractor has seriously tried to submit a good figure and helped the contractor in the process, he should be given real consideration in working toward an award.

Sharp practice on the contractor's part starts when a new bidder is brought into the picture by the contractor without fair consideration to those who have helped most in securing the contract.

It is naturally somewhat discouraging to subcontractors when they fail to receive awards. The typical purchasing agent operates in a field in which every agreement he concludes pleases one but must disappoint several others. The subcontractor who has been accorded courteous treatment and who knows that the subcontract award was fair and square maintains his respect for the contractor despite his disappointment. If he feels that he was given a brush-off or that unfair favoritism was involved, the negotiation could reflect on both the purchasing agent and his company when the matter is discussed by those affected.

It is interesting to note that there are two schools of thought among subcontractors regarding field coordination and contractor's supervision as it affects their prices. The contractor who notifies his subcontractors that the concreting work must be poured on a specific date may annoy some of the "field team." From the management point of view, however, it is only by such advance scheduling and an ability to overcome obstacles as they arise that the job can go forward to assure profitable operation for all concerned.

CHECKLISTS

It is quite conceivable that study of the project since the proposal was submitted has resulted in finding different methods whose use should be reflected in the actual subcontracts. For instance, an overtime program on rock excavation and an expedited schedule on structural-steel deliveries may now be obtainable and beneficial, provided the trades coming on the job subsequently would still be able to meet the advanced schedule. This could result in an earlier completion of the project and net saving in cost.

Occasionally, isolated trades can be put on an overtime schedule to accomplish some particular objective. Usually, however, the workers in other trades will become less productive unless they too are given overtime. This, of course, can become extremely expensive when deadlines are involved near the completion of the job.

This points up the desirability of saving time at the start of the project by the judicious use of overtime for fewer men to expedite their particular phase with minimum expense. Drafting work for structural-steel shop drawings and overtime use of excavating equipment are two examples that come to mind.

A speedup program involving some overtime should only be entered into when it is apparent that the time gained will not be wasted by lack of architectural or engineering information or by trades unable to accelerate sufficiently to take advantage of the better timing.

One of the helpful "props" in negotiating is the purchasing man's cost record book (Fig. 8) in which breakdowns of contract prices, showing quantities and unit prices, have been listed for the various trades on different jobs. The best time for securing quantity information is at the time of negotiation. The data need rechecking by the contractor to verify that the information is realistic and not weighted to provide for an initial greater return to the subcontractor rather than for payments proportioned uniformly with progress. This quantity information should not be used to release the subcontractor from furnishing the actual requirements of the job.

THE NEGOTIATOR

If the contractor's chief executive does not personally handle subcontracting, all negotiations should be routed through one individual who should be responsible for the results. It is never good policy for a group or committee to negotiate contracts. This does not mean, of course, that the contractor's field or engineering staff should not be consulted whenever necessary by the man doing the purchasing.

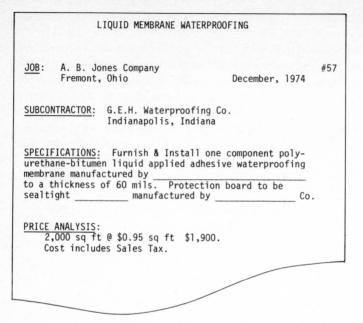

LIQUID MEMBRANE WATERPROOFING

<u>JOB</u>: A. B. Jones Company   #57
        Fremont, Ohio   December, 1974

<u>SUBCONTRACTOR</u>: G.E.H. Waterproofing Co.
                    Indianapolis, Indiana

<u>SPECIFICATIONS</u>: Furnish & Install one component poly-
urethane-bitumen liquid applied adhesive waterproofing
membrane manufactured by _____
to a thickness of 60 mils. Protection board to be
sealtight _____ manufactured by _____ Co.

<u>PRICE ANALYSIS</u>:
        2,000 sq ft @ $0.95 sq ft $1,900.
        Cost includes Sales Tax.

FIG. 8   Cost Record Sheet.

It is quite natural for the contractor himself, who is most deeply interested financially in the success of the company, to become involved in discussions with major subcontractors. Similarly, even though subcontractors may know that their contact should be with the purchasing man, there frequently may be a tendency to offer a few thoughts to the contractor personally, in the hope that it may influence the decision on the award. A very close working relationship between the contractor and his purchasing man is therefore essential. New information, price reductions, withdrawals of bids, and countless other changes can come with great rapidity. The "boss" and his purchasing man need to operate on the same wavelength, or confusion and loss of opportunities for better buying can easily result.

PREPARATIONS FOR NEGOTIATING

Just as the period when bids were solicited was active, there is great activity in the contractor's office after the public announcement is made that he has been awarded the general contract. As the subcontractors bring out their proposals, new alternatives or changes may need to be priced. The contractor must move forward with all deliberate speed to refamiliarize himself with all the bids and to bring the general superintendent or someone charged with field operations into the picture to express ideas on working conditions.

Since the proposal was first submitted it is quite possible that additional bids have been received from subcontractors who may have bid to other contractors. Some bids may have attractive features. Possibly there are combinations of bids or separate awards that could generate savings. Sometimes obvious errors of the subcontractor or contractor come to light that should be resolved before contract negotiations proceed.

Frequently points will be raised in some bids that could affect pricing in other proposals. In putting the estimate together earlier it may have been necessary to act only on the basis of a particular subcontractor's estimate without reference to other bids.

BONDS AND BONDING

Where general contracts require performance and payment bonds (Figs. 14 and 15)—which is usually the case on public work—the prudent contractor will decide on similar protection from his subcontractors. When it is not a contract requirement, quite frequently contractors, especially those in the private, nonspeculative building field, will prefer to deal with subcontractors whose financial stability is not subject to question and will decide not to ask for a bond.

On the other hand, if a particular subcontractor is selected for an award and it becomes evident that he is bonded to perform most of his other work, the contractor must face the possibility of the subcontractor's default, not only on his own project but also on other bonded jobs, about which he knows little. Should a bankruptcy occur, the bonded jobs would have priority in working out a settlement. Consequently, it would be well to add the cost of a performance and payment bond in comparing this subcontractor with some other qualified bidders.

On large lump-sum jobs, the problems of completing the work are so extensive, even though the subcontractors are bonded, that good judgment would usually dictate this added protection regardless of cost.

If it has been the practice to deal with subcontractors from whom bonds are normally not required, when the occasion does arise, there is need for all parties to become "bond conscious" and to recognize that failure on the contractor's part either to furnish material or services when promised or to make payments without unjustified delays would be grounds for complaint by the subcontractor and excuse by the surety in complying with the subcontractor's obligations.

FIRST NEGOTIATING SESSION

One of the first objectives of most subcontractors during a negotiation with the contractor is to expedite a decision to receive the award. Probably the subcontractor is convinced of the validity of his proposal and is now trying to clinch the award. He prefers a session where final agreements which are not subject to further discounting can be made immediately or soon thereafter.

Presumably it has already been established by previous experience that the subcontractor is acting as a principal and has authority to negotiate, just as the buyer has been given his guidelines and limitations in subcontracting. A business card will indicate the position which the subcontractor's representative holds in his firm and help to establish his authority.

The subcontractor is apt to say flatly that "This is it—I can't cut a nickel" or to indicate that "This is about it," which may turn out to be exactly the case or may represent a starting point from which some further reduction could be made. The contractor, on the other hand, wants to create a possible saving to meet his estimating target; in any event, he wants to avoid an overrun of his cost.

Some subcontractors suddenly will become willing to make a greater reduction because they can see that it will expedite the award decision. A list of possible economies posted by subcontractors to reduce the cost, even if potentially attractive, would require time for the architect's analysis. Some subcontractors will prefer to make the price reduction, still quoting on the basis of the drawings and specifications, moving ahead in the knowledge that they might be recommended for the award. In such cases, there probably was a sufficient allowance in their bid to make the reduction, but there was also a tendency to "sit tight" on the original proposal amount.

This is about the time to confirm the "rock bottom price"; and a good lead question—"What do you think it can be done for?"—is an icebreaker in getting negotiations underway.

One subcontractor's common expression after a token reduction is, "This is about my best figure, but please call me before you close." Here we have a case of the subcontractor desiring to stay in the picture and hoping for a final opportunity to talk again about the job. The purchasing man somehow needs to ferret out the accuracy of the comment. Sometimes, he can suggest a very nominal reduction to see whether that amount would establish a final figure or whether a much greater reduction could be anticipated. In any event, before the interview ends, the man doing the buying must be able to feel that the subcontractor has about pinpointed the level below which he could not go.

LEAVING THE DOOR OPEN

At the same time, unless a clear picture has emerged during the interview for the particular trade, an experienced purchasing man will tend to leave the door open for further discussions or suggestions to be volunteered by the subcontractor later. If a bid is hopelessly out of line, the subcontractor should be informed. Otherwise, it is generally advisable in initial discussions not to finalize the bargaining so that the subcontractor closes his file completely on the job.

FACTORS INFLUENCING PRICING

In practically all cases, if the contractor takes time to study similar jobs, he develops a feeling for justifiable costs and for those conditions warranting contingencies. With this study there comes an assurance that is invaluable in negotiating—the assurance of a chap who holds four aces in his hand.

Each project has its differences in the makeup of the subcontract trades and their effect on each other. The one principal factor which has changed since the subcontractors first submitted their bids is that the job has now become an actuality. A noticeable change in thinking usually takes place.

For the general contractor intent on awarding subcontracts, a quick rundown of some conditions that can influence prices in some trades can pay off. For instance:

Excavation Have any dumping areas become available recently which would reduce trucking costs? Could nearby property be purchased and used for disposal of fill, thereby increasing its value?

Structural Steel Do any of the miscellaneous iron items lend themselves to inclusion in the structural-steel work, such as loose or attached lintels or special framing, on which a lower comparable tonnage price could be anticipated? Possibly there may be savings through substitutions involving higher-strength steel (with less tonnage).or heavier rolled sections rather than built-up members requiring more shop fabricating time.

Millwork Can provision be made for more shop fabrication and delivery of larger units to minimize field work?

Glazing Has a basis for replacing broken glass been established?

Toilet Partitions Have alternate prices been obtained from both partition bidders and iron subcontractors for furnishing and installing a hanger system for ceiling-hung partitions?

Regardless of the contractor's preference for allocating subcontract work, having prices from two trades for the same work may offer better buying opportunities. However, it must be recognized that coordination of the work must be considered, as well as the possible disadvantages of fabrication or installation errors if two subcontractors are involved.

WORKING RELATIONSHIPS IN THE FIELD

The working relationship between the contractor's organization, especially the contractor's superintendent, and all subcontractors might be described as a two-way street. It can have a decided influence on the prices which the subcontractors will submit for future work and on the cost of the entire project for the contractor and the subcontractors.

For example, the superintendent recognizes the experience and technical knowledge which a subcontractor has as background in his particular trade; the subcontractor recognizes the talents which the contractor's superintendent possesses for organizing the work. By working together as a team, they obtain maximum efficiency.

Obviously, the superintendent needs to know the drawings and specifications, the gaps, and the interpretations already made, and to have an understanding of the course to follow in clearing up questions.

Following are some characteristics of a "good" superintendent:

■ He should be open-minded when receiving requests for more information or obtaining architects' interpretations of the requirements. He should secure promptly whatever is needed, rather than permit incorrect installation that requires adjustment later or allow work to be held up.

■ He should work closely with the subcontractors in scheduling the commencement of work, delivery of materials, manpower requirements, and presence of proper facilities at the site. He should endeavor to make job conditions favorable to allow subcontractors to perform their work as efficiently and economically as possible.

■ He should inspect the quality of the workmanship and see that proper standards are established as the work starts, letting it be known that inferior work will not be acceptable.

■ He should endeavor to coordinate the work with a minimum of interference with allied trades and permit the work to be completed with the fewest practical return trips; and he should obtain approval of it when it is completed.

■ He should encourage subcontractors' supervisors to visit the site and become familiar with construction underway and its effect on their own future operations.

■ He should have a knack for reconciling differences and arguments that develop when strong-minded individuals assert themselves to the detriment of the effort.

■ He should develop a reputation for fairness and evenhandedness in his dealings with the subcontractors, to make them feel a part of the team working *with* him rather than *for* him.

■ He should be alert to those few subcontractors who, by delaying tactics or mismanagement, are hampering the job's progress; he should secure aid from his superiors in the office who are in a position to give it, in sufficient time to be meaningful.

■ He should be conscientious in acting to process subcontractors' requisitions to avoid delays in scheduled payments.

■ He should take special pains in confirming property survey and bench marks to establish correct axis lines and grades for all trades before some become involved with improper measurements.

The benefits of developing a harmonious working atmosphere on the job cannot be overstated. Building is a complicated business requiring the talents of many trades, the coordination of the labor forces, and the orderly assembling of materials—more often than not—on a congested site. The problems become much less complicated if the work of the various subcontractors is properly coordinated. A reasonable amount of give and take fostered by a knowledgeable superintendent can eliminate petty differences, reduce costs, bring about a speedier completion, and generally enhance the reputations of all parties concerned.

THINGS TO REMEMBER

Aside from the element of unfairness to other bidders, a subcontractor who uses the approach "What do I have to quote to get the job?" often indicates a lack of familiarity with the job because it has not been carefully estimated. In the long run, the subcontractors who stay in business are those who know their costs and who are not tempted to cut further because of information which may be imagined or implied from another's bid. Subcontractors' graveyards are filled with those who relied on another firm's bid—no matter how obtained—rather than on their own appraisals.

It is always good practice to refrain from disclosing information about other bids. Occasionally one may feel inclined to indicate that a bidder is in second or third place without mentioning prices. While negotiations are proceeding, it is best not to indicate that a certain bidder is low, since by the time final prices are received, another firm may have submitted a more-attractive proposal.

Throughout the text, various pricing situations and conditions for awards have been assumed. Actually, depending on the circumstances and the individuals involved, the various possibilities are endless. But let's go back to our original estimating.

THE DEMOLITION AWARD

Only two bids were received, and these varied widely from $4,800 to $8,000. Fortunately the sums involved are not great.

In preparing the estimate, the decision was made that a performance bond would be asked for and included as it would furnish a measure of protection in having the work completed. One of the risks in demolition is that the salvageable items may be disconnected and removed before the time-consuming operations have been completed.

Demolition agreements should state that the title to the salvage does not pass until actual removal from the site. To the extent that labor has been expended in stripping and salvaging, the contractor by retaining ownership builds up a certain reserve for future performance.

Subcontract agreement forms requiring the furnishing of materials and performance of work can be used for demolition subcontracts if appropriate deletions are made in the printed form. If they expect many such jobs, some contractors develop a form for this type of work which simplifies the preparation of agreements and which is more readily acceptable to subcontractors whose work does not involve new construction.

Architects' specifications frequently do not go into much detail in describing demolition, leaving it and its coordination to the discretion of the contractor. Some items are worth consideration by the contractor, for example:

Foundation walls are to be removed to the level of the sloping sidewalk which follows street curb.

Certain usable salvaged materials will be cleaned and turned over to the contractor for his use.

No claim will be made by the subcontractor for light fixtures which were stolen or broken between his inspection of the premises and the date of the subcontract. Sometimes these are of little value before contracting but priceless after an award.

Minor changes can be made in the work without notice to the bonding company. (This would not apply if the contract were doubled in scope, for instance.)

Insurance certificates indicating limits, expiration dates, and 10 days' notice to the contractor in the event of cancellation are to be hand-delivered the following day.

Dumping charges are included in the contract price.

The work will start in 3 days and be completed within 3 weeks thereafter.

Where larger sites and large volumes of debris are involved, consideration should be given to joint operation on the site by the wrecker and excavation subcontractor, possibly even extending to foundation work. Awards for these trades should recognize the effect of involvement with others. These discussions can usually be agreed upon easily during negotiations, but could be grounds for claims for interference, delay, or extra costs if not finalized and if operations do not proceed exactly as planned.

While debris may be left in the basement, a conclusive agreement should not be made until the condition is discussed with the prospective excavation subcontractor and his reaction obtained, because of the additional quantities and types of materials involved.

Provision for obtaining all permits should be included in the demolition subcontract. This may involve cutting of services, street openings and repairs, rodent extermination, etc. The subcontractor is familiar with the routine in securing such permits, and the contractor does well if he does not become involved.

Should a default occur under a bonded demolition subcontract, there is value in including a provision giving the contractor the right, upon 3 days' written notice to the surety, and 20 days or less after the default, to take over the defaulting subcontractor's equipment and award a subcontract for completion. This has the advantage of permitting work to proceed without time-consuming involvement in rebidding to the bonding company.

It is presumed that the demolition work should start promptly. Having satisfied himself that the low bid is not unreasonable, the contractor should conclude the agreement with A at the contract price of $4,800, which included the cost of the performance bond.

To secure the bond, the contractor will have to prepare the contract promptly and have it executed by the subcontractor so that the bonding company may examine the agreement before formally issuing the bond.

THE EXCAVATING AWARD

We recall the uncertainty in connection with low-bidder A whose $20,000 quotation for excavation appeared low when compared to the other bids, as well as when analyzed by using his own deductive unit prices and the contractor's quantities.

The bidder was asked to come to the office to settle contract details. It developed that his estimate had been in error. He showed actual costs of $24,000 and claimed he would lose his shirt at the $20,000 price.

This presents a "What would you do?" question, with the subcontractor's attitude playing a large part in the answer. Assuming that A can demonstrate his estimating error and has not decided to renege completely on the job, the contractor would be well advised to give A the chance to requote. Bidder B's price was $29,000, and C had quoted $32,000. These bidders should be asked to recheck, but we would not expect prices to be in the lower brackets.

The contractor has no other bid close to A, even when A stated that his cost was $24,000. If he is given a new opportunity to adjust his figure, A would probably tend to be appreciative and temper his extra costs somewhat. Perhaps he might even quote $25,000, and the contractor would be left with a subcontractor reasonably satisfied by the price and no saving on the general contract since a contingency of $5,000 had been included in the estimate over A's original $20,000 bid.

Had A stated that he needed $30,000 and would not settle for less, it could be pointed out to him that he has quoted above the next bidder and obviously is out of line. The contractor must decide how far to press his case with the subcontractor to recoup the $5,000 overrun. When the contractor is faced with a loss due to a subcontractor's error, he has the option of trying to buy out from someone else or of getting the "culprit" to stand some part of the loss. Frequently, the second bidder will move up closer to a competitive position, thereby avoiding a suggestion of legal action which would be a considerable nuisance. Normally, the hard-nosed attitude toward a subcontractor who has erred in estimating does not pay off. Responsible subcontractors will try to stand by their proposals even though it represents a hardship. The general contractor must face the fact that a subcontractor who is losing money could be an unwilling member of the project team. This could prove more of a hardship than any financial advantage attained through adherence to the estimate target.

No one relishes a lawsuit to collect for an error of this type. In some cases, but infrequently, it could reflect on the contractor's bidding practices. When putting the proposal together, if the contractor does not see all situations through rose-colored glasses and is not too thoroughly optimistic, the tendency will be to encounter some losses and some savings. A businesslike but understanding attitude when dealing with someone else's errors and a decision to "get on with it" will usually pay off for the contractor.

The case of A's oral bid of $20,000, by telephone, resembles a similar actual situation where an Ohio subcontractor withdrew his bid on which the general contractor had relied. The court of appeals held that the subcontractor had made a clear and definite offer, hoping for the contractor to rely on it. The contractor did rely on it, and it would be unreasonable to permit the defendant to revoke its offer.[1,*]

THE HEATING, VENTILATING, AND AIR-CONDITIONING AWARD

By the time formal negotiations began it is assumed that low-bidder A has found the error in his estimate and submitted an all-inclusive revised bid of $103,000. Though his bid is now in the range of the other bids, his bidding performance did not generate confidence. It seemed advisable to concentrate on negotiations with the other bidders who were asked to contact their major suppliers for best prices.

Bidder D and bidders E and F jointly are close. The contractor foresees no disadvantage in a single subcontract with D as long as reasonable percentages are agreed upon for the carrying charge which D would place on his subcontracts.

The comparison between the revised–all-inclusive and original proposal is noted below:

	Revised		Original	
B	$107,000		$109,000	
C	$106,500		$109,500	
D	$100,000		$106,000	
E	$ 76,000	($102,000)	$ 78,000	($105,000)
F	$ 23,000		$ 23,500	
Alternate #1	$ 3,000		$ 3,500	

Bidder D repeated his preference to omit the concrete bases for equipment to the point where he suggested a heating, ventilating, and air-conditioning final price exclusive of the bases but including the specified temporary heat and maintenance. As the concrete subcontractor had shown a willingness to "contribute" the bases, the heating, ventilating, and air-conditioning trade will show a saving of $1,500 under the original estimate.

Anchor-bolt layouts, templates, and anchor bolts will be supplied under the heating, ventilating, and air-conditioning subcontract for installation by the concrete subcontractor.

Assuming a median cost figure of $4 per square foot for our 25,000-square-foot building, D's proposal of $99,000 would be consistent with current-day expectations. On the above basis, D will be recommended for the work. Subject to securing approvals, a subcontract will be prepared confirming the agreement (Appendix C, Form 1) as well as a

* Superior numbers refer to Notes at end of chapter.

letter of intent (Fig. 10), in both of which bidder D is the heating company.

TIMING AND SCHEDULING OF THE WORK

The bar chart (Fig. 4) referred to earlier can help in discussing the job timing, but it should never be offered as a guaranteed schedule of operation. The subcontractors themselves will want to make their own study to obtain more accurate estimates of when their work will start, based on experience on past jobs and consideration of the available work force.

With wage rates changing frequently, the time schedule becomes critically important. The start of excavation, usually shortly after the award or when the demolition is completed, can be fairly well determined since it primarily depends on moving equipment to the site.

Structural-steel work will start in the field as soon as sufficient steel has been fabricated for continuous erection. Assuming a competent fabricator and no major problems with shop drawings, submissions, approvals, or foundations, this major pacesetter can be expected to conform to the initial schedule.

From this point on, it is sometimes difficult to be precise in establishing work periods unless similar jobs have been performed and schedules recorded. Most general contractors expect their subcontractors to perform their work in accordance with the field construction schedule as it develops. The subcontractor is expected to help himself and his own subcontractors by cooperating and starting as soon as possible with as many men as can be effectively employed.

The contractor should be wary of statements in proposals implying that if the work is performed after a certain date the subcontract price automatically will be increased. It might be that the subcontractor failed to have enough men working at the right times, thereby prolonging the completion. These open-ended statements can be very troublesome when they crop up unexpectedly in proposals which were not studied sufficiently.

In another situation in Nebraska, the subcontractor agreed to start work on erecting steel storage tanks within 3 days after receipt of notice to proceed and to complete the work within 5 weeks after the contractor had completed preliminary foundation work. The foundation work was delayed and was not completed until the date when the entire job had been scheduled for completion. The subcontractor refused to proceed, and another subcontractor had to be engaged.

The supreme court ruled that contract times are usually not "considered to be of the essence" unless there is a specific reference to that effect. The contractor's delay did not relieve the subcontractor of a contractual duty to perform.[2]

It is worth noting that where subcontractors have invested some of their own funds in jobs, there is less likelihood of a failure to start work, as sometimes happens when new conditions arise.

The sequence of work will usually follow the superintendent's instructions, but each subcontractor has a right to expect normal working conditions and reasonable access. A glazier who was led to believe that he could use a rolling scaffold will want either a floor to roll on or plank runways provided by someone else at no expense to him.

Subcontractors often have to be instructed to return to the site to complete their work for any number of reasons. Commonly called "comeback work," it occurs most frequently in the masonry, lathing, and plastering trades. Often it involves tenant work in buildings which was phased for completion several months after the building shell. It may be caused by delays of other trades or by damage to work which requires patching. The condition has to be recognized as a possible source of controversy, and it had best be laid on the table for discussion, with reasonable ground rules set up at the outset. These conditions may be difficult to resolve initially, but at least some agreed-upon terms offer a better basis for later discussion than a claim after the subcontractor has started work and is reluctant to proceed without an exorbitant settlement.

Unlike problems involving defective workmanship, claims for delays are not easily evaluated. Frequently the claimant's attitude may be that the sky is the limit. The failure to deliver and install material of relatively minor value can generate damage claims of thousands of dollars because of delays. Court decisions on delays are not always consistent. Where problems arise, local legal counsel should be obtained.

WHERE FIELD EXPERIENCE HELPS

Despite the contract terms, problems will arise that can take many meetings and diplomacy on the superintendent's part to resolve. Some may be due to personality quirks of foremen or a tendency to handicap others to save face. For example, the steel erector, faced with a shortage of riveters or welders, may attempt to slow down the concrete trade by leaving unriveted or incompletely welded steel. In terms of tonnage, erection progress may look very good, but the job is being handicapped; or by moving or "jumping" the second floor of protection planking to the higher working level, steel erection can proceed, but the concrete subcontractor must slow down because more stories of steel are "open" above and unprotected than the building code permits.

Carpentry foremen may slow up the completion of a particular area of formwork for concrete slabs, making it impossible for the reinforcing crew to complete the work for the next slab pour or for the electrician to install conduit.

The possibility of the various trades' becoming acutely overtime conscious always exists when proper scheduling requires a specific area to be completed and finished periodically. Here is one place where good supervision of field work really pays off by avoiding unnecessary overtime.

If the purchasing man has learned some of these methods through field experience and mentions them in his negotiations, chances are they will not occur on the job. In any event, he has laid the groundwork for a specific complaint with one of the subcontractor's principals, should the occasion arise.

He should have an opinion regarding the coordination problems between trades and should be prepared to see that the areas of conflict are resolved during contracting if the specifications have not already provided for them.

For instance, the tile and terrazzo subcontractors, who normally place their materials on top of the rough concrete floors, should be on notice to verify the concrete work as it is being installed and should be expected to install a greater thickness of fill if necessary because of irregularities in the rough-floor level.

The furring subcontractor should coordinate his work with that of the electrician installing frames for recessed light fixtures. Either could do so if it had been called to his attention previously.

The plastering subcontractor should provide a proper base for following trades such as quick-set tile, and he should avoid damage to adjoining finishes. His men should be on notice to call to the attention of the superintendent those items needing protection (if protection is not included in the plastering subcontract) rather than go blithely on and damage some especially hard-to-replace finish.

Frequently subcontractors will perform for each other without recourse to the contractor. This is usually the more-desirable method as it avoids issuance of work orders that increase the paperwork for the contractor's field office.

Steel-erection subcontractors often hoist metal deck, piping, and mechanical or elevator equipment. To the extent that the contractor is in a better position to bargain for this hoisting, he makes it possible for the various subcontractors to take advantage of this rate schedule as "ceiling prices" and avoid extreme charges for special rigging expenses later.

Under some circumstances, this hoisting could delay steel erection.

In negotiating, the point should be settled with the erector whether the other subcontractors may have to hoist after the normal working day in order to maintain steel-erection schedules.

Another point which should be resolved when dealing with steel erectors or other "heavy" subcontractors would be the number of cranes or derricks which the subcontractor intends to use and the time they will be operating on the job. There are times, for instance, when better progress can be made if an additional crane is supplied for unloading and handling steel. The erector naturally wants to utilize his equipment full-time. As a concession, when negotiating, it may be possible to secure a promise of more equipment when the job reaches a certain stage or time. If there is no discussion, there may be stalling or even complete "reneging" on equipping the job.

SOME IDEAS FOR THE PURCHASING AGENT

Most businesses have sales quotas. There are occasions when an extra contract can help a monthly, yearly, or even a lifetime quota that the subcontractor's representative especially wants to better. Substantial price concessions can be made if the general contractor's buyer is familiar with the conditions and can take advantage of them.

Some jobs carry a certain prestige, and subcontractors may particularly desire to be associated with them—to the point where some additional price concessions will be given. Perhaps competitors have had an exclusive representation in the territory, and another reputable company desires to break into this new market. Perhaps some manufacturer particularly wants to promote a new product and is willing to underwrite part of the cost on this project to get it started.

Finally, in comparing some types of purchases, structural steel for instance, a record of unit prices on recent jobs can be helpful in avoiding major overpricing.

As a rule, steel prices can be expected to be reasonably consistent in any given territory for similar type buildings. When a considerable variation occurs, it could be the result either of a change in market conditions or of inflated pricing. In both cases, investigation is necessary. If one could work back to the pricing of an earlier job and receive the bidder's justification for the market change, the difference, if any, would represent an inflated price which could be subject to further negotiation.

Estimates on the mechanical and electrical equipment, especially in the preliminary stages, frequently may be prepared without a thorough study of true manufacturing costs by referring to price catalogs. Once it has been settled that the job is moving ahead, most manufacturers will be anxious to have a chance to recheck their costs for that particular job and offer lower prices through their local sales representatives or through increasing discounts previously quoted.

A contractor with sufficient volume of work should be able to combine his various projects to secure better pricing and generate savings on many items such as large orders of metal flooring or roof deck.

Some manufacturers have more or less standard-size equipment units. If a 75,000-pound-per-hour steam generating unit is required, one manufacturer may make this size regularly, whereas the next-nearest bidder might have a 95,000-pound-per-hour unit.

Initial costs and operating expense on the larger unit would justify a careful comparison and chance of negotiation, to say nothing of the need to check with the engineering department regarding comparative efficiency and desired performance.

Some subcontractors with large shop facilities are faced with heavy overhead charges ranging up to 75 percent or more of the shop labor costs. These charges, regardless of the amount, are a true factor in operating the business, and to ignore them would bring financial disaster.

At the same time, the percentages can vary widely, depending on the yearly volume, and can be grounds for considerable price adjustment under some circumstances.

Most proposals, especially if firm lump-sum prices were not requested, are apt to include escalation clauses to allow for increases in wage rates and material costs.

Generally speaking, it is advisable to try to negotiate for a firm price to avoid controversies because of the extent of labor expended in the early or later stages of the job. This usually depends on the cooperation of the subcontractor in properly manning the job. Escalation, not to exceed a specified sum, is another alternative.

There is merit in having the initiative for a lower price come from the bidder rather than through urgings on the contractor's part. The subcontractor who takes a subcontract at a price he has suggested is not in a good position to complain as the job moves along, if unforeseen conditions arise that become troublesome to him. At the same time, and aside from pricing, there may be more-advantageous terms or conditions to advance progress which the subcontractor would willingly accept if they are brought out in advance by the contractor. Most subcontractors would be cooperative in accommodating themselves to slight revisions without appreciable additional cost at this period of negotiation.

Expediting piping tests can be helpful, especially when approaching cold weather could delay the tests and the closing in of shafts. A week's delay in this respect could cause an extended deferment because of weather conditions in Northern area jobs and added work in draining systems to avoid freeze-ups. Sometimes, including test valves in addition to those shown on the drawings will permit partial testing by zones and will release other trades in the process.

It always helps to get a building enclosed. Using 2-ply partial roofing on the job in the rainy season, preliminary to the final installation, or promptly glazing windows on a Northern job may avoid layoffs because of bad weather and save several working days in the process. Getting the exterior doors hung will help make the building more secure.

The contractor can consider himself fortunate if there is at least one subcontractor "pushing" and willing to spend some overtime to maintain schedules. A typical example, and one worth aiming for, is to have the electrician include overtime for conduit installation to help regularize progress on pouring concrete, thereby allowing better coordination of all trades.

Percentages for overhead and profit on extra work and for lower percentages on work which the subcontractor himself sublets should be established, while negotiations are still in a "fluid state." A standardized form applying to each general contract can serve as a check sheet in arriving at percentages for the usual type changes, and later it can be incorporated into the subcontract (Fig. 9).

Experience may suggest revisions which could be followed on other contracts. It may also be possible to agree on somewhat more favorable terms than many trade associations offer as standard. These points are entirely subject to bargaining if they are agreed upon as part of the contract negotiation.

Prompt awards may make it possible to combine to make larger orders or possibly to place orders for foreign manufacturers' products still with sufficient time to meet the schedule of the project in question. Occasionally orders are canceled, and a manufacturer may need new orders to fill in the resulting gap in his production schedule.

Frequently there are advantages in combining the work in two sections of the specifications—excavation and foundations, as an example, or curtain walls and glazing. This tends to avoid disputes over responsibility if interferences occur or defective work appears.

In the latter case, for example, the subcontractor with the larger portion of the work would normally expect to be paid a carrying charge if he takes over the glazing contract. It can happen that both subcontractors are willing to team up under one agreement, but the holder of the larger subcontract would expect at least 5 percent, if not 10 percent, for the assumption of the risk. On the other hand, the problem may not involve the matter of working under one agreement so much as the financing which the second subcontractor requires. If the general

TRUST COMPANY
Columbus, Ohio
Contract #207

Procedure applicable to Establishment of Costs covering added
or omitted work from the scope of the original requirements

1. Lump Sum

   Predetermined lump-sum additions to and omissions from the
subcontract are to be based on the following:

Labor: Estimated cost of direct field labor including payroll
         taxes, insurance, assessments, and fringe benefit costs
         and the like, plus overhead and profit at __% for addi-
         tions and at __% for omissions. (Percentages include
         use of tools, supervision, items of cost not heretofore
         mentioned, field and office overhead, and profit.)

Materials: Estimated net cost delivered at the site including
            all applicable taxes plus overhead and profit at __%
            for additions and at __% for omissions.

Work Sublet by Subcontractor: Actual cost to subcontractor plus
            overhead and profit at __% for additions and __% for
            omissions. These percentages apply to the net difference
            in the requirements for additions and omissions.
            Additions and/or omissions shall be itemized in a form
            satisfactory to the general contractor and so as to per-
            mit ready analysis and evaluation.

2. Time and Material

   The cost of work additional to the original subcontract requirements and authorized to be performed on a Time and Material
basis is to be established as follows:

Labor: Cost of direct field labor including payroll taxes, in-
         surances, assessments, and fringe benefit costs and the
         like, plus __% for overhead and profit. Percentage is to
         include use of tools, supervision, items of cost not here-
         tofore mentioned, field and office overhead, and profit.

Materials: Net cost delivered at site including all applicable
            taxes plus __% for overhead and profit.

Work Sublet by Subcontractor: Actual cost to subcontractor plus
__% for overhead and profit.

FIG. 9   Establishment of Costs Form.

contractor agrees to pay the sub-subcontractor (the firm with the smaller
subcontract) directly, there usually is no problem or extra expense.

The contractor must recognize that there will be times when it is
impossible to secure price concessions that are needed to avoid a loss.
Some study may result in changes, combinations, or special terms which
represent concessions of value, and yet it will not secure the price reduc-

tion which the subcontractor, for some reason, hesitates to give. The task for the contractor is to work out a solution that satisfies both parties and still avoids an increase in the price which the subcontractor would normally ask.

For instance, in working out details, the subcontractor or vendor might agree to conditions such as the following:

1. The 10-month rental period for furnishing and installing a sidewalk bridge might be extended to 1 year without additional cost.

2. An additional sliding-gate assembly for truck entrances in the temporary fencing at the site could be provided without additional cost.

3. The usual cash discount of 25 cents per cubic yard for ready-mixed-concrete payments by the 10th of the month following delivery could be deducted from the cost per yard and arrangements made to pay semimonthly by the 10th and 25th of each month.

4. The demolition subcontractor might agree to turn over without charge certain usable planks which he will be salvaging from the job during his work.

5. The hollow-metal-work subcontractor, having ample storage facilities, agrees to manufacture and properly store the job's requirements, thereby making it possible to avoid escalation on shop manufacture which would have been incurred by later manufacture.

6. Most projects will require occasional temporary hookups by mechanical and electrical subcontractors or may need conversion from temporary to permanent service. This must usually be done without interference to other trades. The work itself can be estimated, and the subcontractor could be expected to perform it without claiming overtime expenses if it is brought up during the buy out.

7. Some leeway in equipment-rental agreements is usually obtainable if considered at the time of negotiation. When equipment is being rented on the basis of a 22-day month, rainy days or holidays sometimes can be credited against overtime usage. Rentals can normally be credited against an option to buy and arrangements made for crediting excessive breakdown allowances against rental payments.

FINAL NEGOTIATING MEETING

This is the time when it helps to be ready to buy—to have tied up any loose ends with respect to other bidders and to have confidence that the matters entering into the buyer's analysis are, in fact, trustworthy. Uncertainties, guesses, and gambles should have been cleared up by this point.

Experience with subcontractors in negotiating brings with it a realization that there are other considerations besides price that must be evaluated. Naturally, a proposal that keeps within the contractor's estimate may be attractive, but further details such as the ability to produce a workmanlike job on time are important. His ability to handle the job and to avoid labor complications or shortages all need study. Here the experienced contractor has some advantage in being able to refer to previous contracts and note the subcontractor's abilities or shortcomings. Some latitude in evaluation by a contractor's superintendents and office staff is understandable, but if all records are consistently good or consistently bad, it is worth recognizing at the time of negotiation.

Finally the discussions about the trade are complete, and subcontractor A has been selected to perform the excavation work for $25,000. The contractor is satisfied that this represents the most advantageous arrangement which will at the same time be within the original estimate.

Before reaching the stage of writing a "request for approval," usually directed to the architect or owner, is a good time to verify the points that have been discussed with the prospective subcontractor including possible changes that may have come to mind in recent days. The list would include agreement on items such as the following:

1. Verification of firm price—not subject to escalation. If escalation is permitted, are charges imminent? If so, and the manufacturer has knowledge which he withholds from the contractor, the result will be a feeling on the contractor's part that he has been taken in, to say the least.

2. Definite conformance to drawings and specifications or exceptions, if any. Normally most sections of the architect's specifications refer to the general conditions of the work which are to be considered a part of the trade section. To the extent that these conditions provide for the performance of work by one or more of the subcontractors, their inclusion in the subcontract bids should be confirmed so that the contractor will not have to absorb any costs or responsibilities.

3. Waive "fine print" of the proposal or with certain exceptions.

4. Type of subcontract form including insurance provisions.

5. Insurance to cover explosion, collapse, and damage to underground utilities (XCU), although not specified, will be provided without additional cost.

6. The price on the option for certain alternates will apply for, say, 30 days.

7. Hourly, weekly, and monthly rental rates for equipment which may be used on extra work.

8. Rates for time and material work.

9. Submission of special sketches, if any, including follow-up for building-department approval.

10. Procurement of all special permits in connection with the work.

11. Will attend meeting to resolve joint operation with demolition subcontractor. (Buyer should remain involved during the preliminaries of the job.)

12. Will furnish additional financial guarantees, if warranted by examination of financial statement to be submitted.

13. A bond will or will not be required to be furnished by the subcontractor. The cost would be borne by the contractor if a bond is deemed advisable, but the subcontractor should be able, and would agree, to secure the bond.

14. Any agreement is dependent upon approval of the contractor's recommendation by the architect and owner.

15. The proposed award is to be considered completely confidential at this stage and until specifically released by the purchasing agent.

This final discussion, frequently a telephone call to the subcontractor before the award, is the one opportunity to tie up loose ends that may have been revealed after discussion with other bidders. Infrequently, it may even bring about a decision that there is not a meeting of the minds between contractor and subcontractor. It is far better at this stage to thrash out the details before an award, or even decide not to contract, than to get involved in a time-consuming controversy later.

Despite the fact that the bidders know they should be dealing with the contractor, there are occasions when they will make contact with the architect or owner directly, perhaps offering concessions that could have changed the buyer's analysis. The moral is to keep in touch with all bidders during the buying period to avoid any embarrassment that can come through surprises.

APPROVAL LETTERS

Most specifications, even on a lump-sum contract, require that the contractor is to notify the architect before awarding a subcontract. In some cases this extends further, to the actual approval of a particular subcontractor. Subject to the procedures agreed upon, it is well to submit such a letter promptly.

There are times when bidders who believe they are not likely to be successful will attempt by various means to change a recommendation. Therefore, until an award is definitely approved, it is well not to reveal publicly the direction that is being followed. Once a decision on an award is made and the results publicized, the unsuccessful subcontractors will appreciate the chance to talk about their proposal and get a slant on the market conditions for future bidding, even though their bid was unsuccessful at the time.

SUBCONTRACT FORMS

Further details of the actual preparation of the subcontract forms (Fig. 19 and Appendix C, Form 1) are given in Chapter 10. It is presumed that all bidders will have been made aware of the form to be used, but in discussions with subcontractors and the actual typing of the agreements, the purchasing agent should be careful in defining the prosecution of the work.

It should be agreed, for instance, that the subcontractor shall carefully coordinate his work with job requirements and shall furnish at all times sufficient materials, skilled workmen and equipment to perform the work in accordance with the schedule as it develops and to the entire satisfaction of the contractor and so as not to delay the whole or any part of the work (Appendix C, Form 1, Art. III).

But it is well not to restate subcontract provisions among other points. The questions then could arise, "Why is it being done? Is it intended to imply an extra significance or to downgrade those provisions which are not repeated?"

An important point is that the subcontractor should not employ men, means, materials, or equipment which may cause strikes, stoppages, or disturbances by workmen employed by the contractor or other subcontractors in connection with the project.

Percentages for overhead and profit on changes should be agreed upon and reflected in the subcontract. This may be done by incorporating the establishment of costs form (Fig. 9) into the subcontract.

The owner normally carries fire insurance on the building and materials, but the subcontractor is expected to carry coverage on tools, equipment, and personal property used by the subcontractor or anyone employed by it in the performance of the work.

Finally, the subcontractor's unemployment insurance number for the state where the work is being performed should be noted, together with the state license number if this is required. Otherwise there could be a legal inability to enforce an agreement with an unlicensed subcontractor. These numbers are usually filled in by the subcontractor when executing the agreement.

Usually two copies of the subcontract form are prepared for execution, with additional copies made for those requiring the information in the contractor's, architect's, or owner's office.

Contracts should be prepared promptly and submitted to the subcontractors. It is well to stress the need for prompt execution and return. There are decided advantages in maintaining acceptance of the printed form by all subcontractors. Once it is known in the trade that exceptions are granted easily, there may be a tendency to request many changes and to quibble over details. This can generally be avoided if examination of the form was provided for in the bidding information and if the form was accepted previously on other projects.

If a change obviously becomes necessary, the contractor should agree to have the portion rewritten or arrange for the issuance of a subcontract information letter (Fig. 18), which becomes a part of the contract documents. Subcontractors should be impressed with the necessity for the execution of the contract to permit processing of monthly requisitions and payments to be made under the agreement.

FINANCIAL STATEMENTS

Contractors and subcontractors who have been in the habit of working together in their local community for a period of time are apt to be acquainted with each other, but even in these cases it is well to request an up-to-date statement when major contracts are being undertaken.

An audited statement, prepared in accordance with the special variations for the construction trade, is quite essential whenever parties are negotiating for the first time.

There can always be exceptions to recommended procedures. For instance, some thoroughly qualified subcontractors will lack credit or financing for the purchase of materials. After careful consideration, the contractor's personal knowledge of their reliability may fully warrant an exception.

Most subcontracts and payment bonds provide for payments to be made after materials have been delivered and installed. Frequently subcontractors will request that payments be made for materials delivered but not installed.

Where the furnishing of materials without job-site labor is involved, it is customary to authorize by means of a purchase order (Fig. 25). Changes are confirmed by a modification order (Fig. 26). Aside from the terms of the payment bonds which would require approval if exceptions are made, when payment for material is made as a routine practice, sooner or later confusion can result, and a careless contractor may find that his payments have been used to finance some of the subcontractor's other projects.

CONTRACT PREPARATION

While many contracts have been started on the basis of verbal understandings, the practice can never be recommended. One of the parties may die, or conditions may alter circumstances for one of them, making it less desirable to proceed.

Just as the contractor will find it necessary to secure some written authorization before making commitments, subcontractors will feel the same urgency in confirming the basis of their authorization before becoming deeply involved in ordering materials or taking on new obligations.

In Chapter 8 examples are given which bear on the point of whether or not agreements were fully consummated. Suffice to say at this point that a brief letter of intent (Fig. 10) in connection with an award

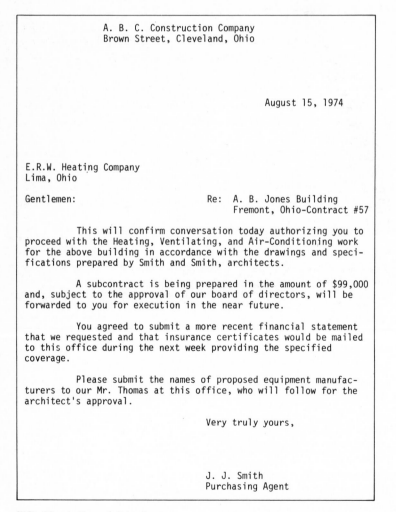

A. B. C. Construction Company
Brown Street, Cleveland, Ohio

August 15, 1974

E.R.W. Heating Company
Lima, Ohio

Gentlemen: Re: A. B. Jones Building
                                         Fremont, Ohio-Contract #57

        This will confirm conversation today authorizing you to proceed with the Heating, Ventilating, and Air-Conditioning work for the above building in accordance with the drawings and specifications prepared by Smith and Smith, architects.

        A subcontract is being prepared in the amount of $99,000 and, subject to the approval of our board of directors, will be forwarded to you for execution in the near future.

        You agreed to submit a more recent financial statement that we requested and that insurance certificates would be mailed to this office during the next week providing the specified coverage.

        Please submit the names of proposed equipment manufacturers to our Mr. Thomas at this office, who will follow for the architect's approval.

                          Very truly yours,

                          J. J. Smith
                          Purchasing Agent

FIG. 10   Letter of Intent.

gives a solid basis for the proper drafting of an agreement, helps toward a clearer understanding, and expedites the initial stages of construction.

Care in drafting the agreement pays off. It helps to dictate contracts while matters are fresh in the purchasing man's mind, but he must be sure that, in the desire for conciseness, a word or a phrase has not been omitted that could cause controversy later.

The innocuous phrase in a pipe-covering subcontract that "all concealed heating risers and radiator connections shall be covered" is an excellent example of a troublemaker. In this case, the dispute centered on whether exposed and concealed radiator connections were both to be covered. It was ultimately held that the word "concealed" applied to both risers and radiator connections, but a suit was necessary to resolve the ambiguity.[3]

The frequently used expression that items are to be delivered "knocked down" needs further description regarding the extent to which some shop assembly may or may not be performed. In the case of a television tower, several weeks after the contract was signed, the contractor furnished the steel erector with detailed erection drawings as agreed. These drawings showed assembly bolts to form the composite members and erection bolts to attach the gusset plates. It had been agreed that the erector was not to perform any fabrication. The dispute in this case revolved around whether the erector was to install 12,000 erection bolts called for on the contract drawings or 50,000 total as actually required. The trouble would have been avoided if the statement "Make all connections, whether or not shown on the drawings" had been included in the bidding documents.[4]

Needless to say, the words "practicable" or "reasonable" do not always mean the same thing to both parties. The case is cited of an engineer who agreed to design an addition to a plant for a charge of 8 percent of the construction cost. Before proceeding with the design, it was agreed that the engineer should check the adequacy of the present structure to determine whether a second story might safely and "practicably" be added. For this study, a charge of $250 would be made. The investigation was completed, and it was determined that it was physically "practicable" to proceed with the proposed construction.

The engineer then filed the required structural plans with the building department and obtained approval. When bids were taken, the lowest was $35,000. The client decided not to build the addition, claiming that it was not "practicable" since he had not contemplated spending more than $20,000. He refused to pay the engineer more than the $250 for the preliminary investigation.

The letter agreement which had been executed was ambiguous in not defining "practicable." Because the intention of the parties was not obvious, oral evidence was admitted to explain the meaning as interpreted by the parties. Ultimately, the jury accepted the engineer's contention and reasoning that the $250 covered only the preliminary investigation. The authors point out that because of the ambiguity, oral evidence was permitted which ordinarily would not have been in the case of a properly worded agreement.[5]

Both parties in a negotiation should make notes of the clauses, terms, and conditions that need to be set forth in the final agreement to be sure that later interpretations will be in agreement. Sometimes there will be missing links when discussing involved clauses, which can be inserted at a time when both parties are in a cooperative negotiating mood. It is a good idea to date all such notes as well as to indicate the time when many discussions take place.

The need for some overtime should be recognized. If the subcontractor is responsible, he should be expected to perform this at his own expense. If the contractor authorizes overtime with the intention of accepting the additional cost, under some contracts he would pay the cost without overhead and profit; or perhaps it would be agreed that limited overtime costs would be borne by the subcontractor. However, the subcontract should be specific regarding the intent.

It should be noted that the percentages for insurance and fringe benefit payments for straight-time work differ from those for overtime work. This would normally be verified when submitting invoices.

If the job is being operated under a "wrap up" insurance program (all subcontractors insured under an owner's program), most of the subcontractors' premium costs should be excluded and necessary adjustments made in the contract price.

Care should be taken that unit prices and alternate prices quoted by the subcontractors are based on the same data as the contractor used in his proposal to the owner. Occasionally, low bids are accompanied by high unit prices or alternates, and adjustments may be necessary to establish the correct relationship with respect to the contractor's agreement with the owner.

When work is performed for nonprofit or public organizations, exemption certificates will be furnished through the owner, and taxes should not be included in the subcontract amount. In such cases, common practice calls for the owner to pay the material invoices directly, following submission and approval of the subcontractor's requisition.

It should be recognized that during the final negotiation, the submission of an approval letter, and the preparation of a subcontract, there may be new facts or ideas that arise. They should be acted upon and agreement reached immediately so that there will be a consistency and acceptance of the final details of a mutual understanding.

NOTES—Chapter 4

1. *Wargo Builders, Inc. v. Douglas L. Cox Plumbing & Heating, Inc.*, 268 N.E.2d 597.
2. *Kingery Construction Co. v. Scherbarth Welding, Inc.*, 185 N.W.2d 857 (1971).
3. *Wood & Company v. Alvord & Swift*, 232 App. Div. 603, 251 N.Y.S. 35; *aff'd* 258 N.Y. 611 (N.Y. 1932).
4. N. Walker and T. K. Rohdenburg, *Legal Pitfalls in Architecture, Engineering, and Building Construction*, McGraw-Hill, New York, 1968, pp. 39–41.
5. *Jovis v. Charben, Inc.*, 199 N.Y.S.2d (N.Y. 1960), in ibid., pp. 9–10.

CHAPTER FIVE

A Cost-Plus General Contract: Preparation of Subcontracts and Items Handled Directly by the Contractor

On the basis of the assumptions outlined in Chapter 3, we are taking it for granted that the owner has decided to select a general contractor to perform the work under a cost-plus-fixed-fee contract and is particularly anxious to get the work started promptly.

From the contractor's point of view, while the profit realized may be less than through a lump-sum contract, the risks are substantially less, and unforeseen conditions in the alteration work will not eat up the anticipated profit. At the same time, both the contractor and his cost-plus subcontractors must not lose sight of the need to perform efficiently to retain their knowledge of the market and working conditions.

The usual method of handling cost-plus work either by a general contractor or his subcontractors is based on one of the following:

1. Cost of the work plus an agreed percentage for profit or fee
2. Cost of the work plus an agreed fixed fee for profit
3. Cost of the work plus a fixed fee with a guaranteed total cost including fee

PROS AND CONS OF COST-PLUS WORK

In the first case the client pays the entire cost and, if a reasonable percentage is agreed upon, the method is frequently used for complicated work which does not lend itself to an accurate definition of scope. A disadvantage from the client's viewpoint is the greater the cost, the greater the contractor's or subcontractor's profit. The client needs to be "sold" on the contractor's qualifications to avoid later regrets.

In the second case, an agreement can be made on the amount of the fee which would remain fixed for the duration of the project regardless of its final cost. This represents a fair agreement for both contractor and client. Normally, minor changes can be made which might vary the cost without a change in fee. If major increases in scope occur, some adjustment in fee is usually provided for.

Understandably, owners would prefer not only to know the amount of the fixed fee but also to have a guaranteed cost, which brings us to the third basis for handling cost-plus work.

Depending on the status of the working drawings, estimates can be obtained and a mutually agreed-upon figure set as a guaranteed limit which the contractor would be responsible for. In such cases, it is customary to provide some measure of participation in the savings (because of the added risk assumed) if the contractor can complete the work for less than the contract guarantee, subject to the adjustments for changes as they occur during the life of the job.

Contingencies, if they need to be provided for, may be included in arriving at this guaranteed limit of cost, commonly referred to as the "upset price." They can be separated from other costs and would not be included in computing savings participation if the parties agree. With such arrangements in the subcontract, there may be limitations on the total participation in savings, say, on a 25 percent (subcontractor)–75 percent (contractor) basis, with total participation limited to some percentage, say, 25 percent of the original fee.

Two of the principal advantages of the cost-plus method develop through the cooperative efforts of contractors, subcontractors, and architects in working to fulfill the owner's needs within the cost limitations which have been budgeted and to provide a very considerable saving in time of completion.

MOCKUPS

On large jobs, if there is need to move ahead with the construction and the architect's working drawings have not been finalized, it may be desirable to build a model room or mockup, incorporating preliminary ideas.

All owners are not equally adept at visualizing the architect's sketches. The investment in a mockup could be saved many times over by reaching decisions more readily and by helping to establish controlling measurements for windows, enclosures, air outlets, ductwork, piping, and electrical switches and outlets needed by the architect and various subcontractors, as well as for interior and exterior finishes.

Circumstances vary but, generally speaking, items being incorporated into the mockup should be paid for at some agreed-upon cost rather than on a "no-charge" basis, which could influence the selection. After the choice has narrowed down, other price factors, including transportation, can be evaluated before final approval.

It is important for all parties—architect, owner, and cost-plus contractor—to recognize that prices on the material under consideration may be subject to negotiation up to the time when the selection becomes more definite. From that point on there is apt to be little opportunity for any bargaining.

RELATIONS WITH ARCHITECTS

Many contractors, convinced that their source for new work is apt to be through association with architects rather than through the occasional building which an owner may require, are reluctant to offer alternatives or suggestions which may reflect on the architect's prerogatives, even though cost reductions could be made.

The experienced subcontractor is understandably the most qualified individual to evaluate costs and methods in his particular trade, even more so than the manufacturer because he has developed a closer knowledge of the installation problems. If the architect has called in a subcontractor for his design studies, the chances are that most of the areas for savings have at least been considered. If not, there remains a wide area to effect savings if the subcontractors are allowed to offer suggestions. With a cost-plus job, the contractor is in a good position to encourage the submission of alternatives.

On a large project, when the bids are in and efforts are being made to reduce costs, it is quite possible to develop lists of 200 or 300 changes for consideration. Some may not be acceptable because they do not meet design criteria, do not offer sufficient economies, or may increase long-term maintenance and operating costs. However, it is not uncommon for 100 or more acceptable revisions to be found if the subcontractors are given the opportunity to scan the requirements. A 10 percent or 15 percent reduction in cost can result.

Architects knowledgeable about costs are finding it hard to overlook such revisions as long as their basic design criteria are followed. Frequently, it can mean the difference between a project moving ahead or stopping at the drawing board.

CONTRACT AND SUBCONTRACT NEGOTIATIONS

In a similar situation, two general contractors who were both bidders on a large project might be offered an opportunity to have their subcontractors submit deductions for alternate changes that, if acceptable, would establish a revised contract price with the lower bidder to be awarded the general contract. The changes would be submitted confidentially in writing with the understanding that only those acceptable would be considered in reevaluating the bids.

With cost-plus contracting, changes should be suggested in the interests of improvement and/or economy without the implication that they are advantageous to only one party. Negotiations or better pricing by the contractor can and should be pursued just as frequently as for lump-sum contracts.

When subcontract bids are submitted to the contractor for consideration, care must be taken to give each subcontractor the opportunity to discuss his proposal and agree on a final amount. As a result of discussions, some subcontractors may sense that they are not in the contractor's picture and may decide to contact the architect or owner with offers of some substitutions or even a further reduction. The contractor's negotiating position needs to be thoroughly understood and agreed upon by the owner. At some point, the contractor's representative needs to be able to tell the owner unequivocally that the subcontract price quoted was offered as the lowest possible bid if he is to avoid later embarrassment when the architect or owner produces a revised bid.

GETTING STARTED

The temptation—and it is a very real one—is to proceed with placing orders and starting on construction as soon as the specific requirements are known. There must be a thorough comparison of the current costs with the budget estimates, however, if the job is not to get out of hand. Changes are apt to come up—sometimes radically different construction procedures—that must be analyzed with cost information furnished to the owner before final commitments are made.

Some owners, because of their own customer relationships, feel obligated to adopt a very evenhanded policy with respect to awards of subcontract portions of the work. Certainly there should be no objection to permitting customers to bid on the work. The general contractor, acting in the interests of the owner, is in a good position to take sealed bids from two or three subcontractor customers with the advance understanding that the low bidder will be given the work. Occasionally customers may consider that their power of persuasion will overcome resistance to their higher price. It is the owner's prerogative to handle such situations. There can be no legitimate objection, however, to a straight race with the low man to be the winner.

Particularly when dealing with an owner's customer, it should be made clear to the bidder in writing exactly what course will be pursued. Sometimes a distrust of the contractor's and owner's intentions may cause the bidder to feel that his importance has not been recognized. The result could be hurt feelings or some other unfavorable reaction on the part of the bidder.

CONSIDERATIONS BEARING ON COST-PLUS AWARD

Some considerations that can bear on the decision to proceed with a cost-plus award include:

1. The need to maintain plant operations once it is apparent that output could not be transferred to another plant

2. The need to get construction underway before drawings and specifications are complete to have facilities available as soon as possible

3. In a general way, a sequence of alteration operations which could involve increasing the floor area as promptly as possible (100 percent added area): expediting the construction of new toilet rooms, stairways, and elevators; removing old stairways, toilet rooms, and, eventually, the existing elevator; and finally, installing new flooring at old stairwells and elevator shafts

These techniques call for close coordination between the contractor and the production department but tend to minimize interference with production during the construction period.

Where feasible, the contractor will probably find it possible to award some work on a lump-sum-subcontract basis. There may well be a need for cost-plus mechanical and electrical subcontracts, however, and excavation and concrete work may have to be handled by his own workmen.

We are assuming that the contractor has some competent foremen and workmen in several trades and is knowledgeable about concreting operations. While the slab on grade, new second floor, and roof concrete work might be sublet, once a portion of a trade section is undertaken by the contractor, he would generally perform all work of that nature.

With some trades, masonry for example, the work jurisdictionally could be sublet or performed by the mason builder. Once a portion of the work is allocated, jurisdictionally the balance of that trade would be handled in the same way.

For our example, we have assumed that the masonry wall contruction can be sublet to advantage and that the new interior-partition work is of a nature to be handled on a time-and-materials basis. Because there is not a large amount involved, it should be possible to define this work sufficiently so that the masons could quote competitively on it. The entire masonry operation could then be handled as a lump-sum subcontract.

ALLOCATION OF THE WORK

Let's list the various classifications of work and indicate how they might be handled:

Break up concrete floors for new columns	Contractor
Excavate for new footings	Contractor
Underpin adjacent wall	Contractor
Excavate for ground-floor slab	Subcontractor
Concrete footings	Contractor
Ground-floor slab	Contractor
Second-floor and roof concrete	Contractor
Cut and patch roof for columns	Contractor
Erect new steel	Subcontractor
Exterior masonry	Subcontractor
Masonry partitions	Subcontractor
Miscellaneous iron—stairways	Subcontractor
Roofing	Subcontractor
Windows	Subcontractor
Glazing	Subcontractor
Acoustic ceilings	Subcontractor
Resilient flooring	Subcontractor
Painting	Subcontractor
Elevator	Subcontractor
Plumbing	Subcontractor, cost plus, fixed fee
Electrical	Subcontractor, cost plus, fixed fee
Heating, ventilating, and air conditioning	Subcontractor, cost plus, fixed fee

SELECTION OF COST-PLUS SUBCONTRACTORS

When it comes to the selection of the subcontractor who is to be considered to perform on a cost-plus basis, we may assume that the choice has narrowed down to one or two in each trade. As with general construction work, it is well to avoid competition on dollars as the only basis for establishing the fee. The selection can better be made on the basis of experience and qualifications for the particular job.

At the same time, some guidelines on making the comparisons are essential. This can be done by defining the terms of the subcontract and the various interpretations affecting the handling of the work so that the scope is comparable and known to the bidders.

For example, the terms of a typical cost-plus subcontract would state:

The actual cost of the work includes the following items and limitations:

1. *Labor:* Labor expended at the site of the work and listed on job payrolls as necessary for the proper conduct of the work. General foreman at a rate in accordance with trade regulations and working conditions; mechanics, apprentices, helpers, and laborers at established rates of pay with fringe benefits and insurance costs.

2. *Materials:* All materials and supplies, temporary structures, scaffolding and staging, fuel, lubricants, expendable small tools, and local cartage on plant, tools, and equipment furnished under the fixed fee. New items furnished under this heading should be billed at their net cost after deducting all trade and cash discounts. Reusable items for temporary construction should be billed at prices approved by the contractor.

Expendable tools mentioned above include hacksaw blades, twist drills, pipe dies, cutter blades, files, drafting supplies (not instruments), new ladders, and similar items.

The cost of materials and/or supplies is subject to deduction for reasonable salvage or resale value of items not consumed in the work. A careful record of such items should be kept until all items have been removed from the project. All transporation expenses incurred in connection with the delivery of equipment are subject to the receipt of prior written authorization from the contractor.

3. *Subcontracts:* All sums payable by the subcontract under agreements made with the written approval of the contractor.

4. *Rentals:* All rents incurred with the prior written consent of the contractor for real estate required for office, fabrication, or storage purposes or for special equipment.

5. *Traveling Expenses:* All expenses incurred with the production and transportation of materials and/or equipment, subject to the receipt of prior written authorization from the contractor.

6. *Royalties, legal fees, and deposits:* Royalties, permit fees, damages for infringement of patents, and costs of defending suits therefor, all when acting in accordance with architect's or owner's instructions; deposits lost for other reasons than subcontractor's negligence.

7. *Losses and expenses, not compensated by insurance or otherwise:* Items sustained by the subcontractor in connection with the work, provided they have resulted from causes other than the fault or neglect of the subcontractor. Such losses include settlements made with the written consent and approval of the owner. No such losses and expenses should be included in the cost of the work for the purpose of determining the subcontractor's fee.

8. *Cost of premiums:* Premiums for all bonds and insurance which the subcontractor is required by contract documents to purchase and maintain.

9. *Taxes:* Federal, state, municipal, or other taxes directly chargeable to the work which the subcontractor is legally obligated to pay and does pay in connection with this contract. This includes all taxes upon labor performed and materials furnished, including, among others, the sales and use tax, gross receipts, occupational or license taxes, and unemployment, old-age pension, and Social Security taxes on wages included in the actual cost of the work. This heading includes vacation time and/or holiday time required to be paid by the employer under trade agreements.

10. *Limitations:* The actual cost of the work shall not include items which are included in the subcontractor's fixed fee as listed below or any other cost or expense not expressly stated above or approved in writing by the contractor.

FIXED FEE

The fixed fee covers, but is not limited to, the following overhead and profit items and services for which no other charge should be made by the subcontractor:

1. Salaries or other compensation of the subcontractor's officers, executives, general managers, estimators, auditors, accountants, purchasing and contracting agents, and other employees at the subcontractor's principal or branch office, except employees of the subcontractor when engaged at shops or on the road in expediting the production and transportation of materials or equipment for the work.

2. Interest on capital or borrowed funds; dues to trade organizations or other associations of any nature; established main- or branch-office local or long-distance telephone or telegraph service.

3. The direction of subcontractors and supervision of the field work from the subcontractor's main and branch offices; purchase of materials; arranging of contracts, financial reports, and other reports and statements including cost and indicated-outcome reports.

4. Under the fixed fee, the subcontractor agrees to furnish all tools, pipe-cutting and -threading machines, welding machines, burning outfits, drafting instruments, if required, drilling and hammering equipment, and such other tools and equipment as may be needed for the proper operation of the job in the opinion of the contractor. All equipment furnished under this contract shall be in good working order, and the subcontractor agrees to maintain it in such condition during the life of the job at his own expense.[1,*]

* Superior numbers refer to Notes at end of chapter.

INTERPRETATION

The fixed fee provides for performance of the work in accordance with the drawings, specifications, and other contractual information listed.

Necessary elaboration, enlargement, or revisions to permit proper operation of the building, as contemplated by the owner and in accordance with the most modern layouts for the trade affected, shall not form the basis for additional fee; nor will an additional fee be allowed for temporary or maintenance work which may be necessary for the subcontractor to perform during the course of the construction work on the building.

The provision with respect to temporary and maintenance work is intended to be all-inclusive (unless otherwise specified), provided only that these items cover the operations of the contractor and all subcontractors on the job in the normal prosecution of the work. This provision does not contemplate two-shift work or regular Saturday, Sunday, or holiday work, although upon occasion this will be required in order to prosecute the work. The fee is fixed for all work done for the owner's account regardless of the change in volume unless provision to the contrary is made. (For example, a change in fee will be considered if the extent of the subcontract is increased by 25 percent or more.)

OPERATION

1. The subcontractor should submit a job organization chart and progress schedule for discussion and approval by the contractor. Changes in organization should be made in the interest of the work at any time if necessary in the opinion of the contractor.

2. The subcontractor should obtain competitive bids for all materials, equipment, and subcontracts entering into the work and submit them to the contractor with recommendations. An approval of such items, if the purchase is in excess of $1,000, would be secured from the contractor prior to making commitments. In general, all bids for mechanical or electrical equipment should include a statement of time required for shipment of equipment after approval of drawings. The subcontractor is responsible for securing such drawings for approval promptly.

3. The subcontractor should request prior approval from the contractor on all purchases where the subcontractor does not recommend the award to the low bidder.

4. The subcontractor should include all applicable conditions which are part of his subcontract in any subcontract that he awards. Terms of such sub-subcontracts are subject to approval of the contractor.

5. The subcontractor should advise the contractor promptly of any

increase or decrease in the cost due to changes in the work affecting either the orders or subcontracts already placed. The subcontractor should secure approval of such changes from the contractor before authorizing the work to proceed.

6. The subcontractor should make sure that accounting methods, including timekeeping and material checking, are in accordance with instructions received from the contractor. Items requiring prior approval by the contractor shall not be included in monthly statements until written approval has been given by the contractor.

7. The subcontractor should keep cost-control records of his work in a form approved by the contractor and periodically submit reports for all items entering into this subcontract in comparison with estimates.

8. The subcontractor should submit to the contractor terms of employment, working hours, and rates for all employees (other than those covered by trade union and/or governmental regulations) that are subject to the contractor's prior approval. These items shall not be included in the subcontractor's monthly statement of the cost of the work without such approval.

9. The subcontractor should arrange for his executives and officers to attend conferences and consult with the contractor when and where required for the purpose of determining construction methods and coordinating and expediting the work.

10. The subcontractor will be entitled to be reimbursed monthly for a proportionate amount of the fixed fee earned, in addition to payment for the actual cost of the work. For example, 50 percent of the fee may be payable along with the reimbursement for monthly expenditures, and the balance may be payable within 30 days after satisfactory completion of the work and acceptance of the subcontractor's work by the architect and the contractor. Fee payments should be based on the proportional relation that actual costs to date bear to the total estimated cost.

11. The subcontractor may, at his option, submit two monthly statements in view of the fact that all cash and trade discounts are to be credited to the cost of the work. The first, covering discountable invoices, should be submitted on or about the 2d or 3d of each month. Payment for all items properly included will be made by the contractor on or before the 10th day of the same month. A second statement covering all other charges would then be submitted as soon as possible after the first statement. Payment of all charges properly listed on that statement would be made by the contractor on or about the 15th day of the same month.

12. The subcontractor should recognize the relations of trust and confidence established between the owner and himself. He should agree

to use his best skill and judgment and to cooperate with the contractor and the architect in forwarding the interests of the owner on this project. He should agree to furnish sufficient business administration and superintendence and to use every effort to keep on the job at all times an adequate supply of workmen and materials to secure its execution in the soundest and most economical manner consistent with the owner's interests.

13. The contractor reserves the right to have its accountants inspect the subcontractor's books and records for the work, and the subcontractor should agree to make such records readily available upon reasonable request.

The contractor should attempt to establish in his own mind a reasonable fee for the subcontract work. His own experience will be a gauge. If the same fee is being considered for each subcontractor, the comparison then comes down to their qualifications, the supervision that would be given, and their abilities to produce the job most economically and satisfactorily for the owner. Only by discussion and investigation can it be determined if each subcontractor contemplates offering the same services.

ADDITIONAL POINTS FOR DISCUSSION

1. Would some additional fee be payable if the building is increased in size? Presumably, yes.

2. Would some additional fee be payable if the building, as originally contemplated, cost more than expected? Presumably, no.

3. Would some additional fee be payable if expensive systems were added to carry out the building function as defined at the time of contracting? Presumably, no.

4. Would some additional fee be payable if expensive systems were added beyond those originally estimated? Presumably, yes.

5. Would some additional fee be payable if the job took much longer than originally contemplated? Presumably, no.

Some explanations or examples written into the guaranteed-cost-fixed-fee subcontract frequently can spell out the terms for a better mutual understanding.

For instance, on a substantial heating, ventilating, and air-conditioning subcontract, involving tenant work, the fee was to remain unchanged regardless of:

a. Sequence of operation.

b. Exact dates when information was released for construction.

c. Final completion for the subcontractor's phase of the operation.

d. The magnitude of the subcontract, subject only to the fact that if new items of scope were authorized, the fixed fee would be increased by 10 percent of the approved estimated cost of such additional item. For example, if a snow-melting system were added to the job, the guaranteed estimated cost and the fixed fee would be increased.

The following changes are examples of items which would increase the guaranteed cost of the work but would not affect the fixed fee:

a. Additional items representing substitutions for alternate installations

b. Increase or decrease in the estimated individual material or equipment items

c. Repairs or replacement of existing work as a result of connections to same or because of the development of the job

d. Cutover or connection work to permit noninterruption with existing facilities or changeover from temporary to permanent services

Additional fee would be paid for the performance of tenant work involving the completion of heating, ventilating, and air-conditioning work on certain floors, when and if released. Payment to be made on the basis of the approved estimated costs plus 10 percent of same as additional fee.

PURCHASING MATERIALS

As the alteration job lends itself to the performance of certain work by the contractor rather than through subcontracting, we will discuss steps and assumptions in securing prices from vendors as well as in placing orders.

Ready-Mixed Concrete

Written inquiries to dealers listing applicable specifications, quantities, and strengths required and requesting proposals with all special terms indicated will provide a basis for discussion.

The proposals received will undoubtedly include printed clauses which warrant study such as:

1. Minimum load per truck or price variations for small loads

2. Waiting-time charge if truck is held for unloading in excess of (usually) 20 minutes at the job

3. Deliveries to be made between 8:00 A.M. and 4:30 P.M.

4. Additional cost for adding various admixtures per cubic yard

5. Additional cost for winter-weather charges per cubic yard, in a manner to comply with specification requirements for area

6. Additional cost for use of high-early-strength cement per cubic yard

Usually the length of haul, traffic conditions, and timing of requirements will have an effect on the coordination of deliveries. The dealer should be notified of scheduled major pouring operations well in advance of needs.   From the contractor's point of view, there should be an avoidance of waiting-time charges as well as of delays to the concreting crew.   This can be done if sufficient trucks are available and work is properly scheduled.

There are wide variations in the qualifications of ready-mix dealers, with respect to operating plant equipment and trucks and having the technical knowledge for producing durable concrete.   The contractor will be well advised to visit the plants of possible suppliers to inspect equipment and storage facilities, satisfying himself regarding the ability to meet the standards required.   The American Concrete Institute, as a national organization, has developed concreting information and standards that all purchasing agents should be familiar with.[2]   Bad concrete in buildings which deteriorates because of improper production or disintegrating roads and sidewalks reflect no credit on the general contractor involved.

Reinforcing Steel

Written inquiries to dealers listing approximate quantities, applicable specifications, and tensile strengths and requesting proposals with discount terms will form a basis for discussion.

Most dealers will quote an additional price per hundredweight for the preparation of placing drawings and a lump-sum price for reinforcing accessories such as slab and beam spacers and high and low chairs, as well as unit prices for these items.   Local conditions may determine the tonnage delivered on each truck, with the vendor usually desiring to make the maximum delivery.

The condition of the job, reliability of the source of supply, availability of storage space, avoidance of rehandling, and the need for time to bend steel in the field are among the factors which will influence deliveries; the possibility of factory bending and tagging of steel to eliminate field bending will be affected by jurisdictional working rules.

Some building departments will require certificates indicating mill reports and conformance to specifications.   These should be submitted currently and payment of invoices conditioned upon receipt in proper form.

Lumber

Buying lumber, like coal, is different from buying most commodities; no two shipments will ever be exactly alike.   The cheapest and best lumber for a given purpose in Missouri might not be at all economical for the same purpose in Massachusetts.

The different grades of lumber shade off into one another.   Unless the purchasing man is a lumber specialist, he runs the risk of outsmarting himself if he relies only on his own intuition. There are responsible lumber salesmen in every area whose advice and suggestions can be very helpful. If they are treated fairly, they can be expected to reciprocate and pass along information that will prove worthwhile.

What is available, delivery-time requirements, possible advantages of carload shipments, most economical shipping lengths with due regard to the job's requirements, and special fire-retardant provisions are all factors that can enter into a lumber purchase.

When it comes to a larger project, there are advantages in securing prices on a lumber list, giving approximate quantities of the various sizes and grades. Usually these quantities can be varied if the total quantity is approximately the same. Dealers are understandably reluctant to maintain price protection for the life of a long-term job. Six months' protection is desirable, if it can be negotiated with the understanding that prices may be subject to adjustment at the end of the period. If lumber deliveries are of satisfactory quality and if escalation at the end of the period is no greater than the recognized market changes, most contractors would recommend extension of such an order if the dealer has cooperated in servicing the job.

There are some dealers specializing in items such as plywood, scaffolding planks, fireproofed wood, and similar products used on construction jobs from whom material could be purchased separately, rather than in combination in one lumber order. Because of variations in the job requirements and frequent changes, it is usually good practice to deal with only one lumber supplier at a time for the general items on the lumber list.

There are no hard and fast rules with respect to lumber purchasing. The availability of a reliable dealer; the granting of 10 percent or 15 percent trade discounts for large-volume orders in some areas, if settled promptly; and a willingness to extend credit under certain circumstances will always be factors to be considered when determining lumber awards.

One will usually find when buying lumber that the prices vary—i.e., no dealer is consistently low for each item. From a job-operating standpoint, once a determination is made and the dealer selected, all items should be processed through this firm for the particular lot or job. Where there is a long list of items, it will save time in buying if the dealers are asked to extend their quotations so that the lowest quotation can be selected most readily. For instance, if tabulated:

#1 fir s4s			Board feet	Dealer A		Dealer B	
100	3 × 4	12 ft	1,200	$150/1,000 bd ft	$180	$160/1,000 bd ft	$192
400	2 × 6	12 ft	4,800	$160/1,000 bd ft	768	$150/1,000 bd ft	720
					$948		$912

Cash discounts, commonly 2 percent if paid by the 10th of the month following delivery, represent a valuable saving if the contractor's financial resources can handle these payments or if his overall standing allows utilization of short-term loans to make payment.

Brick and Concrete Blocks

The masonry work on our assumed job is such that a subcontract seems advisable. The purchasing agent will usually become involved and should be knowledgeable with respect to the materials proposed for use. This may involve samples of brick to indicate matching to existing work, samples of concrete block, accessories, etc., which should be submitted for approval by the architect.

If brick from a new source of supply is required, absorption tests may be necessary. With comparatively small orders of brick, attention needs to be given to ordering sufficient on the last load to avoid shortages caused by unusual waste factors or chipping. These factors, of course, are analyzed by the masonry superintendent who should be responsible for the actual delivery instructions in accordance with his own needs.

Brick samples commonly are mounted on panels to afford the architect a chance to select the color and texture desired. It is always well to have some actual, full-sized bricks submitted to verify the size and composition (e.g., solid or cored) if other criteria besides color could affect the selection.

PURCHASING FOR LARGER PROJECTS

The examples given are typical of the types of materials purchased when the major part of the building is being subcontracted. The picture changes radically when the contract involves larger operations with little subcontracting and when the entire range of purchasing materials is encountered because of the nature and location of the project. Work performed outside the United States is commonly in this category as are larger domestic jobs involving considerable buying of operating supplies and equipment. Arrangements may be made by the field organization for erection and installation with local labor, while the home-office buyer will contract only for the supervision by the subcontractor. It is necessary to have an active expediting section to coordinate deliveries for such offices.

Portable batching plants with ready-mix-concrete trucks, to be bought or leased, may be needed for concreting operations. Cement, sand, and coarse aggregates to be shipped by rail, barge, or truck may now constitute a major part of the purchasing man's commodity purchases as the job starts. Housing, canteen service, temporary roads, and railroad connections are just a sampling of the variety of negotiations to be handled.

The site may require large storm sewers and drainage lines which must be installed before much building can proceed. Assuming that suitable aggregates are available, the concrete-pipe manufacturer will study comparable costs on setting up a local casting plant against transporting from other areas. The period of the year and weather conditions could affect both manufacturing and installation procedures.

The need to "get started" should not eliminate the more-important checks on sources for aggregates in the locality unless they have been thoroughly tested previously and proper mixes designed. If existing ready-mix plants are available, the chances are that suitable concrete will be produced, but the "human factor" in omitting cement or doubling the quantities of admixtures is a possibility that should cause all field people to be on the alert. Sand banks or coarse aggregates that "look good" aren't necessarily so, and tests should always be made when new sources are being used.

Cement continues to be one of the basic ingredients in the building process, and the company representatives still make the rounds of the contractors' offices. They tend to have a broad kowledge of projected building construction and its status as well as a general knowledge of contractors and subcontractors in the community. By offering technical assistance occasionally to architects, they sometimes gain advance knowledge about new jobs. An acquaintance with salesmen representing reputable manufacturers can prove a beneficial source of information that keeps the buyer abreast of industry changes and market conditions.

One needs to recognize that there are variations in the qualities of cement—some barely passing muster—and manufacturing failures occur, so that tests and mill reports are always advantageous on jobs of some magnitude to forestall serious situations which could occur when there are failures.

It is not uncommon for the contractor to request his ready-mix-concrete supplier to place orders with cement company A for the quantity of cement which would be used on a particular project. Cement companies usually appreciate this recognition by their contractor customers. If the job is large and storage-bin capacity is available at the ready-mix plant, it may be feasible to request the use of a certain brand of cement exclusively on the project, which permits an even greater means of quality control.

THE PURCHASING FUNCTION

The buying procedures in such long-term cases tend to become more routine and typical of procurement for manufacturing and industrial departments. The preparation of orders becomes more standardized

with purchasing handbooks offering many suggestions. These departmental forms vary for the different companies, but all can serve as examples to choose from.[3]

Most forms require revisions occasionally to incorporate new requirements which have become advisable. Printers should be asked to hold
"mats" to minimize new typesetting, if revisions are likely to be called
for. Since there will probably be increasing use of computer techniques
during the coming decade, thought should be given to development
of forms adaptable to computer processing.

The obvious points such as vendor's name; billing and shipping addresses of the contractor; price, trade, and cash discounts; and terms
of payment are usually followed by date or time for submission of
shop drawings (if required), with shipment to follow within an agreed-
upon period after approval of the drawings. Naturally there should
be agreement about transportation charges with the variations that are
discussed later.

Sales or use tax regulations frequently vary and should be checked
for the particular state, municipality, or agency involved. For instance,
sales tax is usually not chargeable on transportation costs. Material
or equipment purchases, therefore, if made f.o.b. job site, should have
transportation charges indicated separately or at least handled on the
basis of cost f.o.b. plant, with freight allowed, so that the smaller amount
will be taxable.

Sidewalk bridges and hoist towers used for temporary installations
similarly may have lower percentage tax rates than the area standard
for new work. Sometimes, taxes will apply to rentals only, and the
labor portion will be nontaxable.

Cartage charges for rubbish removal should be broken down to permit
tax exemption on chauffeur's expense where the regulations allow.

JUSTIFYING THE HIGHER PRICE

There are times when the price that certain products can be bought
for, cement for instance, isn't the sole criterion when buying; when
special storage arrangements are required, they could cost more than
the possible savings through immediate purchasing.

Perhaps the need for developing an alternate source of supply for
aggregates makes it desirable to have a second producer, even though
it will cost more.

Finally, knowing the lead time for various products, from date of
contract to delivery may make it possible to obtain earlier delivery;
warehouse steel is a good example.

It should be noted in buying steel that heavier sections may have to be purchased because the designed member is not in stock or that longer lengths must be bought and the "crop ends" scrapped. This would not apply to deliveries made directly from the mill but has to be evaluated for a true comparison.

OPEN ORDERS

From time to time, it will be necessary to take unusual steps to obtain materials, supplies, or equipment critically needed for a job.

Most suppliers will expect a purchase order or written communication before executing a request if it is of some consequence. Avoid the practice of writing an open order to "advise price," especially if a good business relationship has not been established. In a sense, this represents a blank check and could cause problems if there is incompetence on the part of the supplier. It is far better to telephone and agree on a price, or at least on a "not to exceed" basis, before authorizing shipment.

F.O.B. POINTS

On all purchase orders, the method of inland transportation should be specified, i.e., truck, railroad, aircraft, or other designations for lighter material such as parcel post, special delivery, special handling, etc. It is necessary to understand the variations in the free on board designations because they determine when ownership passes from seller to buyer as well as who pays the transportation charges. The f.o.b. point should show which of the seller's plants the shipment comes from.

The more frequent f.o.b. terms and their effect on ownership are:

1. "DELIVERED." This means that the seller will pay transportation charges and that title does not pass to the buyer until he receives the materials.

2. "COLLECT." This indicates that the purchase was made f.o.b. shipping point and that the buyer will pay transportation charges. Ownership is vested in the buyer as soon as the goods are delivered to a common carrier.

3. "PREPAID." Unless otherwise qualified, this indicates that the purchase was made on a delivered basis and that the seller will pay transportation charges. Ownership remains with the seller until materials are delivered to the destination specified.

4. "FREIGHT ALLOWED." This may also be stated more explicitly as "f.o.b. shipping point, freight allowed to destination." Either form indicates that the title passes as soon as the goods are delivered to a common carrier but that the seller will reimburse the buyer for the transportation charges.

5. "F.O.B. SHIPPING POINT, FREIGHT PREPAID." More explicitly this is "f.o.b. shipping point, freight prepaid to destination." The effect of this condition is the same as "freight allowed" except that the seller prepays transportation charges. Prepayment of freight in such instances saves clerical work for both buyer and seller and is usually mutually acceptable.

The matter of title to a subcontractor's purchase is of special importance when determining ownership of equipment or materials which are in transit to building sites where the subcontractor may have defaulted.

Many organizations require multiple copies of purchase orders to serve their various departments. It is good practice to show quantities both in numerals and in words if the typing is indistinct. Similarly, abbreviations which may be construed differently (such as car. for cartons or carloads) should be avoided on the off-chance that someone along the line will misinterpret.

It is good policy to state specifically that orders are confirmations, if that is the case, to avoid duplicate shipments.

Throughout the initial period of getting orders processed, it is likely that some employees may not have had experience in the department. Close supervision is required to maintain a consistency of policy and procedures. The chief buyer needs to "live" with his job to correctly gauge the urgency of the delivery requirements and to process the requisitions and orders as needed. At the same time, he must be alert to possible improvements in buying provisions so that all buyers will be utilizing the agreed-upon methods for future purchases.

There is need to develop a circulating "reading file" of correspondence which buyers should have knowledge of, without undue emphasis on its preparation. Lists of requisitions, vendors, order numbers, and promised delivery dates are usually prepared on a daily basis for the information of those concerned.

NOTES—Chapter 5

1. The stipulation for having the subcontractor responsible for repairs to keep equipment in working order avoids controversies over payment for expensive repair parts or for leasing inferior equipment which might be repaired at the job's expense.

    Electric drills and similar items needed for the work would be under tighter control and more apt to be available when needed if the subcontractor has the responsibility to provide them under his fee.

2. American Concrete Institute, *Specifications for Structural Concrete for Buildings* (ACI 301–72), Detroit, Michigan.

3. George W. Aljian, *Purchasing Handbook*, 3d ed., McGraw-Hill, New York, 1973.

PART THREE

Construction Contracting Guidelines

CHAPTER SIX

Coordination with Field Operations

Once a contract has been awarded, the wheels begin to turn more quickly. The responsibility for the management of the job in the office and the field has been delegated.

FIRST MEETING

The initial meeting of the people who have been and who will be associated with the job should be held to set up the organization. Those who have had a part in the estimating decisions and in the contacts with the owner and architect need to be sure that the individuals who will be involved in the future management of the work will be as knowledgeable about the award as those who were involved in its primary phases.

Included in the listing of headings to be noted in the record of the first meeting are:

1. Name and address of the owner and his representative.
2. Date and form of contract.
3. Name of the project with address, mention of size and number of stories, and brief physical description. Job telephone number.

4. Condition of site.

5. Names and addresses of architect and engineer and their field representatives.

6. The exact scope of the work as listed on contract drawings and specifications.

7. Special requests by the owner that would affect the contractor's procedures.

8. Status of building permit. Has the architect submitted drawings for municipal approval?

9. Scheduled completion date; damages, if any, for failure to complete on schedule.

10. Special conditions affecting adjoining buildings.

11. Special conditions affecting streets and utilities.

12. Status of structural engineer's excavation and foundation drawings.

13. Site survey; name and address of surveyor engaged by owner to establish property lines and bench marks for site.

14. Dates when demolition and excavation should start in field.

15. Conditions affecting immediate award.

16. Subcontracts already negotiated, contingent upon the award of the general contract.

17. Insurance coverage and certificates for owner.

18. Need, if any, for priority in awarding structural-steel and mechanical and electrical subcontracts.

19. Contractor's organization.

20. Work to be performed by others.

Minutes of this first and the later meetings should be distributed to others in the company who are responsible for the management, insurance, accounting, timekeeping, payroll and cost control, local banking arrangements, safety, and public relations, when these functions are handled by different individuals.

This initial set of minutes (some headings will be dropped after the first meeting) is sometimes used as the contractor's authorization to his organization to proceed with construction. In a sense it represents a confirmation, usually signed by the contractor and/or his contracting engineer, that the details of contract execution and the client's financial responsibility are satisfactory.

In the case of public works for which bond issues have been approved, there should be a clear understanding of the terms, especially with respect to extra work. A Wisconsin contractor on a municipal facility for which he had a $50,000 contract was later authorized to proceed with a $100,000 extra for which no competitive bids were taken or public approval of a revised bond issue affirmed.

The supreme court held that contracts for extra work made in violation of law are void and unenforceable. Therefore the contractor could not recover on the basis of the contract price or under the theories of equitable estoppel or promissory estoppel, either of which would prevent the municipal agency from denying the validity of the contract.

Recovery could be allowed on the basis that, barring fraud, there should not be unjust enrichment on the part of the agency. No profit would be allowed the contractor, and the final amount of the extra would not exceed that which would have been paid by unit prices under the contract.[1],*

The first meeting for project should only be scheduled when the contract has been secured and—this is important—when the owner's finances have been verified to the point where payment is assured to the contractor for work performed.

Copies of the minutes of the meetings can be useful if they are also sent to the architect and owner as a diplomatic means of pointing out that architectural information is needed on certain items or that a decision by the owner is of importance if field work is not to be delayed. Frequently both architect and owner are invited to attend the meetings to help bring about the liaison needed in coordinating the job.

No one's feelings will be hurt if the two headings Information Needed from Architect and Decisions to Be Made by Owner appear regularly. It will be a source of real satisfaction to all if there are no listings.

Following the general items, it is customary to add the various trades listed in the contract item list (Fig. 7), developed from specifications, so that status reports can be made at future meetings. This list provides a schedule to develop lead times for issuing subcontracts and material requisitions needed for timely fabrication and delivery to the job. The dates when the various items or trades are required on the job (r.o.j.) can be indicated in the minutes of subsequent meetings.

JOB ORGANIZATION

Assignments of the field superintendent, his field assistants, the job accountant (if this function is not handled through the main office), and/or the timekeeper, as well as of individuals in the office assigned to purchasing and liaison with the architect, should be listed.

It is not unusual for many types of buildings to have 30 or 40 percent of the costs allocated to mechanical and electrical features. This requires not only experienced subcontractors in these fields but also trained contractor's personnel who can coordinate the work. This involves seeing that the necessary shop drawings, schedules, sketches, cuts, and

---

* Superior numbers refer to Notes at end of chapter.

samples are submitted for approval, returned, and resubmitted if necessary before fabrication can begin. This work can be handled either through the field or office organization, but it is essential that all orders placed by the subcontractors be listed by someone on the contractor's staff and their progress followed carefully until delivery is made.

Periodic meetings of the various piping and electrical trades serve to bring their progress into focus and also point up areas requiring field coordination of the trades to avoid interferences.

On our proposed building, the field superintendent would undoubtedly handle all trades and schedule meetings to resolve conflicts.

Provision for the following temporary items needs to be discussed and arranged:

> Electric light and power service
> Heat and water supply
> Job fencing to protect against having an "attractive nuisance"
> Plant and equipment which the contractor may furnish from his own shop
> Watching service after completion of demolition and excavation work
> Freight address and most expeditious routing to site

PROCUREMENT ON THE JOB

In smaller construction companies some of the functions in the field or office will be performed by the same individuals. Sometimes illness requires a realignment of jobs, so that a familiarity with the overall operation can be extremely useful. For younger men, still undecided about the phase of the business which appeals most, the different jobs serve as the training grounds and can provide a perspective on the management of the company.

To the extent that routine purchasing work can be performed by a junior engineer in the field office, it not only permits the home-office buyer to concentrate on more critical awards but also gives others a taste of buying and, to a certain extent, of the bargaining that enters into all procurement.

While the preparation of lumber lists is normally a job function, the actual procurement, especially for initial orders, may be handled by the purchasing man. This includes material for concrete formwork, job offices, fencing, guard rails, signs, planking, protection materials, and temporary shanties, if portable stock sheds are not to be used. The use of rented office trailers should be carefully considered, as in many cases they are of great advantage.

It is not too soon for the contractor to think in terms of fireproofed wood for the shanties. Sooner or later a disastrous fire can cause untold damage and set the job back for months. "Fire consciousness" should be the watchword right from the start; equally so "accident consciousness."

A junior engineer, assigned to the field office, can be helpful in securing prices for lumber, scaffolding, and rough hardware. This permits him to get the feel of expediting deliveries from the different suppliers and vendors in the area.

Hardware and lumber dealers are usually quite aware of the start of new construction in their locality, through the use of F. W. Dodge *Reports* or other trade reports, and can be helpful to the new superintendent unfamiliar with the territory. At the same time, if the superintendent is charged with buying these materials, he should move slowly and retain competitive interest in his work until it is obvious who can best serve the project. Here again the purchasing man can offer guidance and usually come up with a few money-saving suggestions. The superintendent is in the best position to furnish names of new dealers and vendors as well as to make inspections of plants where manufacturing would be required.

OFFICE MEETING

Weekly or biweekly meetings at the main or branch office serve the useful purpose of acquainting all the contractor's staff with the progress in the field and in negotiations with subcontractors. Superintendent's reports should be submitted for each meeting. (Most companies have their superintendents submit daily construction reports on progress and on subcontractors. A sketch of the typical floor plans of the building printed on the reverse side of the superintendent's report can serve for indicating daily progress on major trades to accompany the descriptive information given in the report. This serves as a complete job record that can be an invaluable reference.)

It is particularly important that the meetings be held regularly and that problems of the purchasing man that need resolution be settled. Despite the fact that such meetings take time, they serve as an incentive to have all participants report on their progress in accomplishing what is expected of them. The temptation to postpone meetings because of urgent field conditions needs to be guarded against, especially in the early stages of the job, if subcontracting is to move along properly.

Occasionally, an owner intent on moving the job along will attend office meetings. Usually, he can better understand the contractor's problems and needs if he is made aware of them by knowledge gained through these meetings.

CHANGES AND CHANGE ESTIMATES

Hopefully, no changes have been requested by the owner or architect at the time of the first meeting. When they do come, procedures must be established to secure estimates from subcontractors promptly to be analyzed and forwarded to the architect for decision. Outright additions to the work can usually be determined readily on the basis of the contract terms for extra work. The time-consuming complications start when portions of the contract work are omitted in revised plans or when changes are made necessitating revisions in the construction methods the contractor intended to follow. It is particularly important that a pattern be set for prompt handling of change estimates and that a time schedule be followed for consideration and approval or decision.

The contractor should be careful to avoid claiming that the architect is holding up the work when in fact the delay may have been caused by his and the subcontractors' slow processing of the change estimate.

Practically all specifications require approval of change estimates before revised work can be started. The contractor is frequently on the horns of a dilemma. Although a delay in the decision may put work schedules out of joint, a unilateral decision to proceed runs the risk of later objections to cost and an outright refusal to pay for the change. Before proceeding with a change for which he expects to collect additional payment, the prudent contractor will secure approval from whoever is authorized to give it under the contract.

While unit prices sometimes can and do serve for pricing certain additional work, at other times changes are complicated by revisions that involve negotiations, compromise, and, in some cases, extended explanation to the architect and owner before the change is authorized. The purchasing man can help by sitting in at the meetings with subcontractors to facilitate working out a satisfactory settlement. In all cases where extras or credits are involved, the accepted practice, assuming the change results in additional work, is to deduct the omissions from the additions and pay for the added work only on the basis of the difference or the net change.

On larger projects with changes of considerable magnitude, it becomes necessary to obtain preliminary estimates of the costs from the various subcontractors to acquaint the owner with the approximate additional expense or credit. With experience, the individual assembling the estimate will be able to arrive at a figure more quickly than if formal estimates are submitted. Confirmations, of course, should be obtained. If the owner's approval of the approximate estimate has been secured and if the confirmed estimate when received is in the right range, a basis has been set for the work to proceed. If the final estimate exceeds the original approximation, there is usually an opportunity to reevaluate the need for the change and/or revise the requirements by omitting some of the work.

Each project will need its own procedures. The owner must recognize that his change is delaying the work; the contractor and subcontractors must appreciate that a formal procedure is a necessity if they are to have a firm basis for the collection of costs incurred.

UNIT PRICES

When checking piping and electrical change estimates through unit prices, it is customary to total additional quantities of various pipe sizes from which the omissions in these sizes will be credited. The net change in quantity when multiplied by the applicable contract unit price will establish the value of the material in the particular change involved. Commonly, unit prices are set at the base price for omissions and at the base price plus 10 percent for additions. The 10 percent differential normally covers the contractor's overhead expenses, including estimating, and the processing of orders or change orders.

It is well to keep in mind that this differential cannot always be accepted as valid once a project gets underway and unforeseen factors come to light. For instance, a change in standard lighting fixtures may be extremely expensive or hardly significant depending on whether the basic order has been manufactured. Changes in pipe quantities, if the items are on one supplier's order, may be handled with little cost, whereas if different kinds of pipe are involved, it may require dealing with several vendors.

With some types of work, driving steel pipe piles, for instance, the credits for omitted length may be comparatively small—perhaps less than 50 percent of the unit price for the addition—because of the waste factor, once the piling has been delivered to the site.

Unit prices for electrical work involving conduit installation in the early stages of a long-term job and pulling wires at the completion of the project are examples in which variations due to changing wage rates must be considered in applying unit prices.

Before accepting unit prices for the change, thought should be given not only to the scheduling of the work and the extent of the drafting and/or fabrication, but also to the actual procedures involved in making the change and the necessary qualifications agreed upon.

The unit prices which apply "for the duration" can be arrived at fairly only when it is known what proportion of the work will be performed during the different wage periods. Failing this, consideration should be given to unit pricing based on the scale of wages during the projected contract with limited escalation permitted. This could apply to labor based only on the period when it is estimated the new work will be performed, with no further change in pricing if scheduling conditions vary.

SUBCONTRACTORS' CLAIMS FOR EXTRA WORK

The contractor in many cases will be called upon to pass on the legitimacy of a subcontractor's claims for extra work. As an example, while the specifications may spell out the scope of the work, changed conditions encountered may warrant extra payments to the subcontractor for the owner's account. The contractor then becomes the advocate on the subcontractor's behalf of those claims which warrant additional reimbursement.

For instance, when considering changed conditions involving an excavation subcontract, there is reason to grant them, on the basis that they were not foreseeable by a reasonable contractor, and to reject them, if they were of a type produced by the construction operation or inherent in the environment.[2]

When the condition occurs, the prudent contractor will see that it is observed and analyzed promptly. While it is still a change, it can often be settled promptly and cheaply. Claims that are submitted after the job is completed, with no opportunity for full evaluation, result in endless litigation and mounting costs and are of no benefit to anyone.

These changed conditions in connection with excavation are apt to involve two general types, the first of which is defined as "subsurface or latent physical conditions at the site differing materially from those indicated in the contract."

Examples are: encountering rock in an excavation where none had been indicated in the boring logs furnished to the bidders, encountering substantially more rock than had been indicated in the boring logs, or encountering permafrost or ground water not indicated on the contract drawings.

The second type refers to an unknown physical condition of an unusual nature, differing materially from those ordinarily encountered or recognized as inherent in the contract work.

Thus, when rock outcroppings were visible at the site, the presence of rock in the excavation was not a changed condition, even though the boring logs showed it was some distance below the surface.

Again, where a contractor undertook work in a built-up area, he should have anticipated the presence of underground utility lines even though such lines were not shown on the contract drawings.

These examples are generalities in a sense, but the prudent contractor will consider them "straws in the wind" and recognize that divided courts and reversed decisions sometimes result because of circumstances varying only slightly.

FIELD DISPUTES

When a dispute arises in the field over whether or not certain work is included in the subcontract, the matter should be resolved quickly. Where this is not possible, and it is understood that the subcontractor will make a claim, slips can be submitted to "verify time only." This can forestall later inflated time claims which cannot be easily disputed if the contractor's records are incomplete. These disputes need to be kept "open," discussed at the first opportunity, and resolved before the conclusion of the job.

A cardinal rule for the field superintendent is to keep a daily diary in which the operations of the job are recorded: visitors; inspections; tests; meetings with union officials; arguments with subcontractors; and instructions from the office, owner, or architect all should be noted. When controversies arise, as they invariably do, a good diary not only helps to reconstruct the discussion, but also can be the basis for a proper defense if the matter reaches the litigation stage.

Superintendents should be especially watchful to see that time slips are prepared for all controversial claims. This is quite important when a subcontractor has gone to the other extreme in documenting his version of the matter with minute data that are difficult to disprove. A good set of photographs, properly identified, can be helpful in cases where disputes may develop.

A change-order pad, confirming owner's instructions to the contractor, is a valuable form to use, as is a similar authorization given to subcontractors by the contractor. Unfairness or arbitrary behavior on the part of either contractor or subcontractor will encourage similar action by the other and, in the subcontractor's case, inflated extra claims will have to be resolved.

EXPEDITING DELIVERIES FOR
SUBCONTRACT WORK

Most general construction companies expect their field superintendents to keep close track of material deliveries both for their own needs and for those of the subcontractors. All subcontractors should be made aware of their own responsibilites, but experience indicates that it is risky to assume that the subcontractor will take the same interest in expediting his materials and equipment that the contractor does.

Quite commonly, the individual who negotiated with the subcontractor is in the best position to step into the picture and insist on improved deliveries. The person responsible for awarding subcontracts, and offering the possibility of future awards, usually carries sufficient influence to advance deliveries, provided he has been given sufficient notice of the impending delay. The purchasing agent, normally involved in several jobs, should be in a better position to coordinate deliveries than the superintendent, who at this stage may be having other problems taking his attention.

Occasionally, a telephone call can be made from the field office at exactly the opportune moment to straighten out some situation that can seriously affect the progress of the work. For the most part, however, routine telephone calls to the various subcontractors and vendors soon after the award are more productive. This will ensure that the information has, in fact, been received to prepare shop drawings, to secure purchase order numbers being placed for the major items, and generally to develop a working relationship for the job.

MATERIAL DELIVERIES

One of the attributes of a successfully organized construction operation is efficient handling of deliveries of material and equipment. The sheer logistics of moving large quantities of different types of materials by several transportation means, to arrive when needed without creating storage or rehandling problems, is one of the most important and sometimes frustrating tasks which the contractor faces.

Late deliveries of the ready-mix concrete may necessitate working the concrete gang overtime; the temporary loss of a critical carload of brick can knock the props out from under progress and costs on brickwork; or the breaking of the boom on the crane that was lowering the boiler into the basement can easily require major repairs and slow down the mechanical work that was barely on schedule. (Actually, the fact that construction work *is* completed in the face of the multitude of such problems is a tribute to the resourcefulness of the types of individuals who carry on building work.)

Aside from a very few materials, most building products are custom-made. Delivery problems are increased because of the need for shop drawings and their approval and/or correction. Manufacturing delays and occasionally unacceptable materials inevitably slow down progress on the building. It is a rarity to have a job where all the specially fabricated materials are available in the field before installation starts. Changes may be required as the work proceeds, making some portions of the materials unusable and in addition creating storage and rehandling problems that add considerably to the costs.

Good planning for deliveries (often based on sad experience) starts from the day a subcontract or order is negotiated. The construction schedule (or critical path method on larger jobs) will show when work or materials are required on the site. Working backward, allowances need to be made for manufacturing, for securing raw materials, and for submitting shop drawings and corrections and approvals, as well as for negotiating agreements. Some leeway is possible in the early stages of some jobs, but delays in securing information to conclude subcontracts and other factors beyond the contractor's control make it important to follow all portions of the work closely.

It is standard practice when subcontracting work or buying fabricated materials to agree on a schedule for submission of shop drawings, cuts, sketches, and samples for the architect's approval. Upon their release, the latest drawings and specifications on which to proceed should be given to the subcontractor. He should acknowledge that he has the necessary information and agree to submit his first shop drawings at some specific date. It is discouraging to have the steel contractor announce a week or more after an award that he lacks certain critical information for placing mill orders or preparing shop drawings.

With some trades, miscellaneous iron, for example, prompt submission of a list of required items to be furnished by the subcontractor, including a statement of the proposed number of shop drawings to be submitted, has a way of bringing out into the open any misconceptions individual subcontractors have about the scope of the work. In some cases, where requirements are not too clearly indicated on the drawings, contractors may have to prepare their own miscellaneous iron schedule and secure prices on the basis of the schedule, recognizing that a loosely defined job is apt to be priced on the high side by bidders who recognize the gaps.

Hopefully, the lists submitted by the mechanical, electrical, and piping subcontractors on their equipment will match the lists of the specified manufacturers. Extreme delays can arise in the approval of many substitutions for which new engineering data must be produced or structural changes considered. Lower-priced but acceptable equipment will not necessarily create a price reduction if structural changes and redesign eat up the savings. Clearly, the subcontractor who is able to adhere rigidly to a mechanical specification has a headstart in getting his shop drawings submitted, in getting equipment on the production line, and in getting delivery when needed. On large jobs where many shop drawings are involved, this becomes a critical and continuing problem that ends only when the building is completed and "as built" drawings are submitted.

In one of the first contacts with a subcontractor, it is well to confirm that he actually is proceeding, even though formal orders and subcontracts may have been delayed in preparation. Frequently a letter of intent (Fig. 10) can be issued pending formal execution of the agreement. When formal approval of an award by the architect and owner is necessary, failure to issue a letter of intent sometimes imperils the subcontractor's full commitment to proceed.

Many times, subcontractors fail to start work because a letter of intent has not been issued. Once the contractor has official approval, it is a good idea to confirm for the specific subcontract and avoid any delay in release.

The letter should authorize proceeding, should indicate that a subcontract in a specific amount is being prepared, and should request that insurance certificates be forwarded or hand-delivered if the work is to start immediately. If there is need to establish an actual starting date as a basis for time of completion, the letter or "notice to proceed" should be sent in duplicate with one copy to be acknowledged, dated and returned.

PLANT INSPECTIONS

Telephone calls usually, but not always, will have a way of bringing fabrication progress out into the open. Whether the work is really up to schedule or is being sidetracked for some other job can only be determined positively by visiting the plant.

The contractor's representative on a plant visit will often be working for or with the purchasing man and should be familiar with the shop drawings and approvals. He should know the shop order number of the job and some of the distinctive features. A casual wave of the hand by the shop foreman to indicate that "you can see we're 50 percent fabricated" could mean someone else's job of similar material.

The representative should be aware of the actual need for the material he is following. If his exhortations for deliveries resemble the cry of "Wolf!" too often, his effectiveness can become zero. Close coordination of the actual need in the field with manufacturing production is only possible by advance planning and careful working together of manufacturer and contractor.

The contractor's representative may have to develop the skills of a diplomat. The ability to talk the other fellow's language and explain your own needs can help immensely. A representative should be familiar with trade practices involved, since few manufacturing plants handle their production similarly. The variations in structural-steel fabrication and wood cabinetmaking are as different as night and day.

During the course of a construction engineer's life there are countless opportunities to visit plants manufacturing many types of building products. It is difficult to emphasize too strongly the advantages and importance of visiting such plants if only for a few hours. The experience and assurance that comes from these trips can broaden one's horizons and develop the background that every construction man needs when dealing with subcontractors. If the visits are leisurely and are not connected with a particular project, they can be more informative and of lifelong interest.

There are times when an order in production is moving along nicely, and the welcome mat is out to encourage a visit from the contractor. Frequently the production manager has matters under such good control that progress reports are available which will greatly simplify the inspection.

There are other times, however, when tensions develop if the shop schedules are not being lived up to, and the last thing the production people want is another inspection or more "expediting." Although their attitude may be quite inhospitable, it behooves the contractor's man to keep his cool, to remember what he came for, and to quietly but firmly get his information. He may find that the production information being given appears inaccurate or questionable. Experience will have to determine whether this is the time to bring the matter to a head, whether he should phone his superior, or whether he should leave to recheck his data and return after tempers have cooled.

He must be careful to check his "facts"; any slight error could be seized upon by the manufacturer as grounds for discrediting the entire report and prolonging the controversy rather than hastening delivery.

Some manufacturing plants (especially the larger ones) involved in competitive operations are quite squeamish about their production rates. A too-close or obvious scrutiny can produce an offended and noncooperative attitude which is not conducive to getting the desired information or production.

Nevertheless, subcontracts usually include a statement to the effect that inspections by the contractor are permitted to check on the status of production. The contractor's representative may need to have a tough hide so he won't be browbeaten or embarrassed before he can fulfill his mission.

The right to make shop inspections should be clearly established in the subcontract. With some types of materials, especially those where samples have been submitted for architect's approval, shipment should not be made without such inspection unless it has specifically been waived.

Unacceptable materials or equipment can be a source of much controversy and delay to the job. The best time to insist on corrections is always before shipment has been made.

Larger contracts frequently require full-time inspection services with periodic reports both on production and conformance to specification requirements. In such cases, the procedures become more standardized, and the contractor can spot potential troubles as the reports are received.

A loose-leaf reference book with names, addresses, and telephone numbers of the principal manufacturers; their plant managers and shop foremen; and other personal data can be useful for contacts on future jobs. A few minutes to write a thank you letter for helpful attention given by some individuals in a plant can lay the groundwork for later contacts and assistance. Over a period of time, plant superintendents often become plant managers. Cordiality over the years, with a letter of congratulations when there is an advancement in position, will help create a mutually beneficial relationship between the companies involved.

LETTERS TO SUBCONTRACTORS

Even when work seems to be proceeding normally, it sometimes helps to write a letter to the subcontractors (Fig. 11) to make them more aware of a particular project and to remind them of what they might do to assure satisfaction for an owner. For example, the following letter to all subcontractors on a New Jersey project (fictitious names are used) had the advantage of energizing the subcontractors and of impressing the owner with the contractor's genuine interest in the client's work.

Another letter (Fig. 12) helps to confirm occupancy schedules with subcontractors on a job which will require home-office attention to assure a satisfied client.

COST REPORTS

On the larger jobs, cost studies are essential to keep abreast of the outcome. Comparisons with estimated costs usually take the form of monthly "indicated-outcome reports" which establish the closeness of the actual costs with those estimated and the resulting project earnings.

The purchasing agent is in the best position to be able to forecast the outcome of the negotiations for the various trades. The cost engi-

FIG. 11   Letter to Subcontractors re Workmanship.

Contract #_____

Gentlemen:

You are now proceeding, or soon will be, with the furnishing of materials and field work for the Ajax building.

It seems advisable to notify you that the Ajax Corporation and Smith, Smith and Smith, architects, as well as ourselves, desire a high standard of workmanship which is expected to result in a properly finished and creditable building of which all may be proud.

In your own interest, we request that you recheck the specifications on the work and discuss the requirements with your representative who will actually install or supervise the work in the field.   It is particularly important that all foremen be impressed with the necessity of performing the work acceptably, using approved materials, in order to ensure final payment for the work promptly after completion.   We believe you will agree that the difference in cost between a carefully executed job and a carelessly performed contract is negligible.   All employees should be cautioned against inferior workmanship which might require expensive corrective work later.

We recognize that the officials of your company may find it difficult to visit all the jobs which you undertake.   At the same time, in addition to cautioning your foreman on the caliber of work required, we would like to suggest that a senior representative of your organization visit the site when your work is underway to get an idea of the size of the development which the Ajax Corporation contemplates and to satisfy yourself that your own work is being performed as carefully as you know it should be.

Very truly yours,

---

neer would normally refer to the purchasing agent's records as well as secure advice on contingencies to be provided for.

The change estimate file—whether maintained by the job engineer or the purchasing agent—should also be analyzed to determine additional costs and reflect those already incurred.

Schedule slippage, strikes, bad weather, and failure to receive essential information for construction are among the conditions affecting the results and need to be recognized promptly regardless of whether the job is a lump-sum or cost-plus contract.

Cost records serve two purposes for most jobs, i.e., they are used to recognize trends and (hopefully) bring about improvement and to establish historical records for future use.

At the conclusion of all work when all accounts have been settled,

FIG. 12 Letter to Subcontractors re Progress.

May 3, 197___

Contract #_____

Gentlemen:

You are or shortly will be commencing your work on the fourth floor of the above building. It seems advisable that all subcontractors recognize that the owners expect the work to be completed by June 14th to permit their occupancy and use shortly thereafter.

This represents a "tight" schedule which will require close coordination and cooperation by all to bring about a satisfactory result conforming to the owner's occupancy needs.

Please advise your field representative of the above, and follow the work carefully from your office to ensure completion.

Please have shipping and delivery requirements verified, advising us promptly of any items which will call for special attention in expediting.

Please also confirm with our superintendent, Mr. Smith, the size of the field force needed to comply, and have your job fully manned at all times.

We request that you acknowledge receipt of this letter on or before May 15th.

Very truly yours,

---

a final cost report establishing the true costs of all work with applicable quantities, and including that performed by the contractor's employees, will be invaluable for estimating future work.

NOTES—Chapter 6

1. *Blum v. City of Hillsboro*, 183 N.W.2d 47 (1971).
2. George F. Sowers, "Changed Soil and Rock Conditions in Construction," *Journal of Construction Division: Proceedings of the ASCE*, November 1971.

CHAPTER SEVEN

Financial Problems

Many books could be written about financial troubles that involve construction. In this chapter, the field is being narrowed to the problems affecting contractors and subcontractors. In addition, of course, the owner may delay making vital decisions, the architect may be supercritical in the interpretations of his requirements, the labor representative may be uncooperative, or the government may institute new controls. All are costly and affect both contractors and subcontractors. Still other factors bear on the relationship, including problems caused by bad weather, strikes, or poor estimating. James A. Bourke, vice president of The First National Bank of Chicago, has cited lack of experience in building, lack of balanced managerial experience, and outright incompetence as underlying factors that all contribute to forcing the contractor's national average "after tax profit" down to less than 2 percent.[1,*]

One of the obvious advantages of subcontracting portions of the work is to minimize the amount of the contractor's working capital required to keep projects underway. The subcontractors are primarily responsible for meeting their own payrolls and for financing material requirements to some extent. When these individuals are known to the general

* Superior numbers refer to Notes at end of chapter.

contractor, the risks, if any, can be accurately weighed in advance.   However, other factors unknown at the time a subcontract is awarded may arise which can affect a subcontractor's ability to perform on a particular job.

With new bidders, the risk is greater, and it is increased if the job is in a locality unfamiliar to the contractor.   The problem is to figure out whether a certain subcontractor is financially able to handle a particular job for which he has submitted a bid and is now negotiating for a subcontract.

The guidelines which might help in this determination need to be studied in advance because there will probably be little time to analyze financial statements in detail when the estimate is being put together. Pertinent information can be stored, however, and the danger signals spotted before final commitments are made in the subcontract negotiations.

There is a surprisingly small amount of data published which bear on the financial qualifications of subcontractors.   Undoubtedly much material has been compiled by banks and lending institutions which is available for interoffice use, but the risks of misinterpretation are so great that little attempt is made to publish it.

One exception has been the Robert Morris Associates, who prepare and publish studies of financial statements on nearly 300 lines of business, made possible by data from approximately 33,000 different companies contributed by its membership on an anonymous basis.   In 1972, for the first time, a section was included in the "Statement Studies" devoted to building subcontractors, with reports using a "percentage of completion" basis to promote uniformity.

In some of these trades, statistics on nearly 200 subcontractors have been supplied by the reporting banks and have been totaled with the resulting lists shown as upper-quartile, median, and lower-quartile figures, with respect to various ratios, as well as reports on assets and liabilities.

The figures can serve as general guidelines only, making it essential to stress the Disclaimer Statement at the end of this chapter, on the basis of which permission has been granted to include the tables in this publication.[2]

TABLES

It does not follow that firms placing in the upper quartile and in apparent best financial condition are automatically entirely healthy, nor does it follow that firms placing in the lower quartile should be rejected. Rather, the latter firms need closer scrutiny and possibly advances or other support if a subcontract is to be consummated.   There are conditions where individuals in such firms have the practical knowledge and abilities that are not shared by others, and their skills simply must be retained.

At the same time, the risks with the lower-quartile subcontract firms are obviously greater. The prudent contractor will endeavor to secure further guarantees to ensure performance. This could take the form of personal or corporate guarantees or even of a bond, the cost of which the contractor would be expected to cover.

From the contractor's point of view, studies such as this are the first stage in attempting to reconcile the data of financial statements—frequently not comparable—which throw light on questionable reports.

The Robert Morris Associates' tables give some timely ratios that may be recognized and better understood through reference to definitions appearing in the complete edition of the "Statement Studies."[3]

Analysis of significant ratios for contractors breaks down into three types of relationships, namely: (1) balance-sheet ratios, i.e., current assets compared to current liabilities; (2) operating ratios, i.e., profit compared to contract revenues; and (3) profit-and-loss statements versus balance sheets,[4] i.e., profits compared to net worth.

An inquiry into a subcontractor's financial situation needs to be handled diplomatically. Also, a too-close probing of finances before subcontracting could be unsettling to the subcontractor. This makes it advisable for the contractor to have given thought previously to any points needing clarification.

Experienced bankers, considering loans to subcontractors, can and do scrutinize reports more deeply and will usually be interested in a job progress report on uncompleted work (Fig. 13). Importantly, this

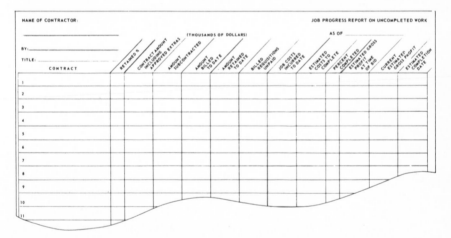

FIG. 13   Contractor's Job Progress Report on Uncompleted Work.

CONTRACTORS—COMMERCIAL CONSTRUCTION

107 statements ended on or about June 30, 1972
241 statements ended on or about December 31, 1972

CONTRACT REVENUE SIZE Number of Statements	Under $1MM 71	$1MM & less than $10MM 220	$10MM & less than $50MM 46	$50MM & over 11	All sizes 348
ASSETS	%	%	%	%	%
Cash	13.9	11.7	7.1	9.2	9.3
A/R-Progress Billings	33.2	27.4	27.3	35.6	30.7
A/R-Current Retainage	8.3	10.3	9.4	9.8	9.8
Unbilled Work in Progress	4.1	7.6	9.4	4.2	6.8
Costs in Excess of Billings	1.3	2.5	3.0	4.0	3.2
Other Current Assets	5.0	5.2	8.2	6.9	6.8
Total Current	65.8	64.7	64.3	69.7	66.5
Equipment	7.8	9.0	10.7	18.2	13.1
Real Estate	3.7	16.5	14.5	4.2	10.8
Joint Ventures	1.5	3.0	.8	2.0	1.9
Other Non-Current	21.1	6.8	9.7	5.9	7.7
Total	100.0	100.0	100.0	100.0	100.0
LIABILITIES					
Notes Payable-Bank	5.8	5.6	6.2	4.5	5.4
Accounts Payable-Trade	27.5	23.1	20.3	18.8	20.6
Accounts Payable-Retainage	2.9	4.9	6.4	4.3	5.1
Federal Income Tax Payable	1.9	3.0	3.3	4.9	3.9
Billings in Excess of Costs	5.2	9.0	4.9	10.8	8.3
Contract Advances	.7	1.0	2.5	.0	1.1
Prov. for Loss on Contr in Prg.	.1	.2	.0	.0	.1
Current Maturities LT Debt	1.3	1.6	2.7	1.2	1.8
Other Current Liabilities	7.8	5.8	5.4	5.2	5.5
LT Total Current	53.2	54.0	51.7	49.8	51.6
LT Liabilities, Unsub.	17.2	16.4	21.1	13.0	16.6
Total Unsubordinated Debt	70.4	70.4	72.8	62.8	68.2
Subordinated Debt	.5	.6	.2	4.2	1.9
Tangible Net Worth	29.1	29.0	26.9	33.0	29.9
Total	100.0	100.0	100.0	100.0	100.0
INCOME DATA					
Contract Revenues	100.0	100.0	100.0	100.0	100.0
Costs of Work Performed	78.9	88.4	88.0	93.9	90.8

CONTRACTORS—GENERAL BUILDING-SINGLE, FAMILY HOUSES & RESIDENTIAL BUILDINGS

107 statements ended on or about June 30, 1972
179 statements ended on or about December 31, 1972

CONTRACT REVENUE SIZE Number of Statements	Under $1MM 120	$1MM & less than $10MM 139	$10MM & less than $50MM 22	$50MM & over	All sizes 286
ASSETS	%	%	%	%	%
Cash	8.6	9.3	6.0		7.5
A/R-Progress Billings	14.2	22.5	34.1		23.2
A/R-Current Retainage	4.1	5.0	9.1		5.7
Unbilled Work in Progress	20.5	13.6	13.9		16.9
Costs in Excess of Billings	3.1	7.1	4.0		13.3
Other Current Assets	10.7	6.1	4.4		5.5
Total Current	61.1	63.7	71.5		72.2
Equipment	10.2	4.9	12.0		7.3
Real Estate	18.1	9.5	8.9		8.4
Joint Ventures	1.0	1.3	1.7		1.3
Other Non-Current	9.6	20.6	5.9		10.8
Total	100.0	100.0	100.0		100.0
LIABILITIES					
Notes Payable-Bank	14.5	11.0	9.3		9.7
Accounts Payable-Trade	14.6	20.6	18.0		15.9
Accounts Payable-Retainage	1.7	1.3	4.5		3.4
Federal Income Tax Payable	1.3	2.1	4.3		3.8
Billings in Excess of Costs	4.4	3.1	2.7		2.9
Contract Advances	4.1	4.2	4.7		2.9
Prov. for Loss on Contr in Prg.	.0	.2	.0		.0
Current Maturities LT Debt	4.0	2.7	2.7		2.4
Other Current Liabilities	8.7	9.9	5.3		5.5
LT Total Current	53.3	55.0	51.7		46.7
LT Liabilities, Unsub.	19.3	18.3	19.9		18.9
Total Unsubordinated Debt	72.6	73.3	71.6		65.6
Subordinated Debt	.7	.4	2.0		3.7
Tangible Net Worth	26.7	26.3	26.5		30.7
Total	100.0	100.0	100.0		100.0
INCOME DATA					
Contract Revenues	100.0	100.0	100.0		100.0
Costs of Work Performed	79.4	85.6	81.1		86.0

		(A)	(B)	(C)	(D)	(E)	(F)	(G)	(H)	(I)
Gross Profit		21.1	11.6	12.0	6.1	9.2	20.6	14.4	18.9	14.0
All Other Expense Net		18.1	8.5	9.6	3.0	6.3	17.4	11.7	15.2	9.4
Profit Before Taxes		3.0	3.1	2.5	3.1	2.9	3.2	2.8	3.6	4.6
RATIOS										
Quick		1.6	1.3	1.0	1.2	1.3	1.3	1.3	1.0	1.3
		1.1	1.1	.9	1.1	1.0	.7	.9	.9	.9
		.7	.9	.7	.7	.8	.2	.3	.6	.3
Current		2.0	1.6	1.4	1.5	1.6	1.9	1.6	1.5	1.7
		1.4	1.3	1.2	1.3	1.3	1.3	1.2	1.2	1.3
		1.1	1.1	1.0	1.1	1.1	1.0	1.1	.9	1.0
Fixed/Worth		.1	.2	.2	.4	.2	.1	.1	.1	.1
		.3	.4	.6	.6	.4	.4	.3	.6	.4
		.6	.7	1.1	1.3	.7	1.0	.9	1.0	.9
Debt/Worth		.9	1.2	1.8	1.3	1.2	.8	1.4	1.6	1.2
		1.7	2.1	3.0	2.5	2.2	2.1	2.6	3.2	2.4
		3.3	3.7	6.0	3.7	3.8	4.7	4.7	6.1	4.7
Revenues/Receivables		**34** 10.6	**39** 9.3	**43** 8.4	**39** 9.2	**38** 9.4	**19** 19.0	**26** 13.9	**35** 10.2	**22** 16.7
		68 5.3	**55** 6.5	**57** 6.3	**55** 6.5	**55** 6.5	**40** 8.9	**48** 7.5	**52** 6.9	**46** 7.9
		106 3.4	**72** 5.0	**77** 4.7	**75** 4.8	**78** 4.6	**100** 3.6	**78** 4.6	**69** 5.2	**82** 4.4
Revenues/Working Capital		12.8	22.5	34.8	31.9	21.5	15.1	22.8	18.0	18.9
		6.0	12.4	15.7	11.2	11.0	5.6	11.7	8.9	9.6
		-46.2	6.0	6.2	6.2	3.9	-70.0	3.6	-47.8	1.0
Revenues/Worth		10.1	14.4	19.5	16.9	14.6	11.3	17.6	16.5	15.2
		6.2	9.2	12.3	9.6	8.8	5.8	11.3	10.9	8.7
		2.7	5.9	6.2	4.4	5.4	2.6	6.7	5.9	4.5
% Profit Before Taxes/Worth		50.0	31.1	30.3	45.5	36.9	47.6	45.3	40.2	45.9
		26.8	16.6	19.8	27.2	18.8	16.7	22.7	26.2	20.8
		-.9	5.4	11.0	10.8	5.9	2.9	8.9	15.4	6.9
% Profit Before Taxes/Tot. Assets		19.0	10.9	7.1	12.7	11.8	14.6	12.6	10.1	13.6
		8.3	5.0	5.5	9.3	5.6	5.4	6.6	5.7	5.9
		-.8	1.7	2.3	3.3	1.8	.8	1.9	2.6	1.6
Contract Revenues		$39851M	$862704M	$1040574M	$1775134M	$3718263M	$58706M	$400199M	$491324M	$1696375M
Total Assets		29533M	405776M	497765M	595512M	1528586M	40702M	181702M	240812M	834297M

SOURCE: Robert Morris Associates. See note 2 at end of chapter.

155

CONTRACTORS—HEAVY CONSTRUCTION, EXCEPT HIGHWAYS & STREETS

43 statements ended on or about June 30, 1972
58 statements ended on or about December 31, 1972

CONTRACT REVENUE SIZE	Under $1MM	$1MM & less than $10MM	$10MM & less than $50MM	$50MM & over	All sizes
Number of Statements	28	57	14		101
ASSETS	%	%	%	%	%
Cash	8.1	11.8	12.6		10.1
A/R-Progress Billings	18.7	28.2	23.3		28.2
A/R-Current Retainage	13.8	7.3	12.0		9.4
Unbilled Work in Progress	2.6	5.4	1.8		2.2
Costs in Excess of Billings	1.3	1.0	3.7		2.3
Other Current Assets	6.2	5.8	2.4		5.4
Total Current	50.6	59.5	55.8		57.6
Equipment	14.1	24.7	30.5		21.0
Real Estate	15.6	2.9	3.8		5.6
Joint Ventures	11.0	.3	3.6		6.4
Other Non-Current	8.6	12.6	6.4		9.4
Total	100.0	100.0	100.0		100.0
LIABILITIES					
Notes Payable-Bank	13.1	5.1	6.6		9.0
Accounts Payable-Trade	12.6	18.8	15.4		18.7
Accounts Payable-Retainage	3.4	.8	2.3		1.2
Federal Income Tax Payable	1.0	3.2	3.7		2.8
Billings in Excess of Costs	9.2	3.1	2.4		5.1
Contract Advances	.1	.3	1.6		.6
Prov. for Loss on Contr in Prg.	.1	.1	.2		.1
Current Maturities LT Debt	2.6	3.3	5.1		3.1
Other Current Liabilities	4.4	7.9	5.7		5.7
Total Current	46.5	42.7	42.9		46.3
LT Liabilities, Unsub.	8.5	14.7	16.7		14.1
Total Unsubordinated Debt	54.9	57.4	59.6		60.4
Subordinated Debt	.3	.6	.3		.3
Tangible Net Worth	44.7	42.0	40.1		39.3
Total	100.0	100.0	100.0		100.0
INCOME DATA					
Contract Revenues	100.0	100.0	100.0		100.0
Costs of Work Performed	67.7	74.7	88.2		86.0

CONTRACTORS—HIGHWAY & STREET CONSTRUCTION, EXCEPT ELEVATED HIGHWAYS

77 statements ended on or about June 30, 1972
166 statements ended on or about December 31, 1972

CONTRACT REVENUE SIZE	Under $1MM	$1MM & less than $10MM	$10MM & less than $50MM	$50MM & over	All sizes
Number of Statements	51	153	34		243
ASSETS	%	%	%	%	%
Cash	10.4	12.7	7.9		8.9
A/R-Progress Billings	16.6	28.2	23.8		25.0
A/R-Current Retainage	4.2	6.5	8.4		7.5
Unbilled Work in Progress	1.9	2.6	1.7		1.7
Costs in Excess of Billings	.2	2.1	3.1		3.0
Other Current Assets	12.5	7.3	7.8		8.4
Total Current	45.9	59.5	52.8		54.5
Equipment	31.1	29.6	33.1		29.0
Real Estate	8.8	3.4	3.7		6.7
Joint Ventures	2.8	.5	.7		
Other Non-Current	11.4	7.0	9.7		9.0
Total	100.0	100.0	100.0		100.0
LIABILITIES					
Notes Payable-Bank	4.7	6.3	6.9		5.9
Accounts Payable-Trade	10.2	18.2	15.2		15.7
Accounts Payable-Retainage	.9	1.0	2.0		1.5
Federal Income Tax Payable	2.8	3.8	3.1		4.1
Billings in Excess of Costs	.7	1.7	1.9		1.5
Contract Advances	.2	.5	1.2		.7
Prov. for Loss on Contr in Prg.	.0	.0	.0		.0
Current Maturities LT Debt	5.8	4.4	4.5		4.9
Other Current Liabilities	6.9	6.7	7.2		7.1
Total Current	32.3	42.6	41.9		41.3
LT Liabilities, Unsub.	14.4	14.6	11.2		13.5
Total Unsubordinated Debt	46.6	57.2	53.2		54.8
Subordinated Debt	9.7	1.7	1.1		2.0
Tangible Net Worth	43.6	41.1	45.7		43.1
Total	100.0	100.0	100.0		100.0
INCOME DATA					
Contract Revenues	100.0	100.0	100.0		100.0
Costs of Work Performed	65.1	79.8	83.2		82.6

	1	2	3	4	5	6	7	8
Gross Profit	17.4	16.8	20.2	34.9	14.0	11.8	25.3	32.3
All Other Expense Net	13.3	13.4	16.0	28.0	10.4	6.2	21.1	36.6
Profit Before Taxes	4.1	3.4	4.2	7.6	3.6	5.6	4.2	-4.4
RATIOS								
Quick	1.5	1.2	1.5	1.8	1.4	1.4	1.4	2.2
	1.0	.9	1.1	.9	1.1	1.1	1.1	1.1
		.7	.8	.7	.8	.9	.9	.7
Current	1.8	1.6	1.8	2.2	1.9	1.6	1.8	2.3
	1.4	1.1	1.4	1.3	1.3	1.3	1.4	1.3
	1.0	.9	1.1	.9	1.1	1.2	1.1	.9
Fixed/Worth	.5	.5	.4	.5	.4	.5	.4	.4
	.8	.9	.7	.8	.6	.8	.6	.7
	1.3	1.4	1.3	1.3	1.1	1.1	1.0	1.4
Debt/Worth	.7	.8	.7	.5	.9	1.3	.9	.7
	1.4	1.4	1.4	1.1	1.5	1.8	1.5	1.4
	2.4	2.5	2.5	1.9	2.5	2.5	2.4	2.2
Revenues/Receivables	35	46	36	30	37	40	36	41
	57	59	56	58	55	55	55	67
	78	84	77	88	84	84	77	97
Revenues/Working Capital	17.9	24.8	18.5	11.0	22.9	17.6	27.8	10.6
	8.9	9.0	10.2	5.0	9.4	14.2	11.9	4.0
	2.6	-50.7	4.5	-28.6	3.8	7.3	5.6	-53.3
Revenues/Worth	8.1	8.9	8.7	5.8	8.1	7.8	8.3	7.0
	5.1	5.1	5.4	4.1	5.9	6.8	6.0	4.3
	3.0	3.4	3.5	1.7	4.0	4.2	4.7	3.1
% Profit Before Taxes/Worth	33.6	25.2	33.0	34.7	42.2	41.1	45.9	26.4
	17.4	15.0	17.7	17.0	19.5	32.7	21.7	5.8
	4.5	4.8	4.2	3.1	4.3	15.2	5.5	-14.8
% Profit Before Taxes/Tot. Assets	13.3	10.9	13.5	14.2	18.0	16.3	21.0	9.9
	7.4	6.8	7.2	8.9	7.4	10.4	8.2	1.8
	2.1	1.3	2.3	1.9	1.8	6.2	2.7	-8.7
Contract Revenues	$1486557M	$630402M	$564263M	$25982M	$666472M	$229077M	$173873M	$14479M
Total Assets	782085M	325165M	278165M	24782M	300496M	105021M	74950M	20999M

SOURCE: Robert Morris Associates. See note 2 at end of chapter.

CONTRACTORS—CONCRETE WORK

9 statements ended on or about June 30, 1972
39 statements ended on or about December 31, 1972

CONTRACT REVENUE SIZE Number of Statements	Under $1MM 21	$1MM & less than $10MM 26	$10MM & less than $50MM	$50MM & over	All sizes 48
ASSETS	%	%	%	%	%
Cash	12.2	8.9			3.3
A/R-Progress Billings	26.4	37.9			19.0
A/R-Current Retainage	4.4	7.0			1.7
Unbilled Work in Progress	.9	3.7			4.0
Costs in Excess of Billings	.2	.2			.7
Other Current Assets	8.0	6.3			1.7
Total Current	52.1	64.4			31.8
Equipment	31.2	22.6			44.6
Real Estate	6.8	4.1			16.8
Joint Ventures	.1	1.9			.4
Other Non-Current	9.9	7.0			6.3
Total	100.0	100.0			100.0
LIABILITIES					
Notes Payable-Bank	9.5	6.3			2.7
Accounts Payable-Trade	13.7	20.7			7.6
Accounts Payable-Retainage	.0	.3			.1
Federal Income Tax Payable	1.2	2.0			1.0
Billings in Excess of Costs	.3	.7			.2
Contract Advances	.0	.0			.0
Prov. for Loss on Contr in Prg.	.0	.6			.1
Current Maturities LT Debt	3.8	6.9			5.8
Other Current Liabilities	5.8	8.1			4.6
Total Current	34.2	45.6			22.0
LT Liabilities, Unsub.	17.7	16.6			22.3
Total Unsubordinated Debt	51.9	62.2			44.3
Subordinated Debt	2.6	1.5			9.7
Tangible Net Worth	45.5	36.3			46.0
Total	100.0	100.0			100.0
INCOME DATA					
Contract Revenues	100.0	100.0			100.0
Costs of Work Performed	74.2	83.8			81.8

CONTRACTORS—ELECTRICAL WORK

87 statements ended on or about June 30, 1972
163 statements ended on or about December 31, 1972

CONTRACT REVENUE SIZE Number of Statements	Under $1MM 125	$1MM & less than $10MM 109	$10MM & less than $50MM 10	$50MM & over	All sizes 250
ASSETS	%	%	%	%	%
Cash	11.5	10.2	12.1		8.5
A/R-Progress Billings	40.9	41.1	42.4		45.7
A/R-Current Retainage	5.7	8.7	5.7		2.4
Unbilled Work in Progress	5.3	5.8	8.6		8.2
Costs in Excess of Billings	2.5	4.6	2.5		4.3
Other Current Assets	13.0	6.5	8.6		8.7
Total Current	79.0	76.8	79.9		78.0
Equipment	9.1	8.7	8.4		8.2
Real Estate	5.5	3.1	5.4		4.4
Joint Ventures	1.1	1.4	.3		2.5
Other Non-Current	5.4	10.0	5.9		6.9
Total	100.0	100.0	100.0		100.0
LIABILITIES					
Notes Payable-Bank	10.7	13.5	5.0		9.8
Accounts Payable-Trade	22.4	15.3	20.0		12.5
Accounts Payable-Retainage	.1				
Federal Income Tax Payable	1.5	2.4	3.0		3.9
Billings in Excess of Costs	3.5	9.7	17.6		7.9
Contract Advances	1.1	.6	1.6		5.3
Prov. for Loss on Contr in Prg.	.0	.1	.0		.0
Current Maturities LT Debt	2.3	1.3	1.3		1.0
Other Current Liabilities	7.9	8.8	12.6		10.5
Total Current	49.4	51.7	60.9		50.9
LT Liabilities, Unsub.	8.0	11.3	3.9		9.0
Total Unsubordinated Debt	57.4	63.0	64.8		59.9
Subordinated Debt	2.0	1.4	.1		8.9
Tangible Net Worth	40.6	35.6	35.0		31.6
Total	100.0	100.0	100.0		100.0
INCOME DATA					
Contract Revenues	100.0	100.0	100.0		100.0
Costs of Work Performed	75.8	82.9	83.5		88.1

Gross Profit	11.9	16.5	17.1	24.2	18.2	16.2	25.8
All Other Expense Net	9.3	11.9	14.1	21.0	12.8	11.5	21.8
Profit Before Taxes	2.6	4.7	3.0	3.4	5.4	4.7	4.0
RATIOS							
Quick	1.8	1.1	1.7	2.0	1.6	1.6	1.4
	1.2	1.1	1.3	1.2	1.1	1.1	1.1
	.9	.7	1.0	.9	.8	.9	.7
Current	2.2	1.5	2.1	2.5	2.1	2.3	1.8
	1.6	1.3	1.6	1.6	1.4	1.3	1.4
	1.2	1.1	1.2	1.2	1.1	1.1	1.0
Fixed/Worth	.1	.3	.1	.1	.3	.3	.5
	.3	.3	.2	.4	.7	.4	.8
	.6	.4	.4	.7	1.5	1.1	1.5
Debt/Worth	.8	1.0	.8	.7	.7	.6	.7
	1.4	2.3	1.4	1.5	1.4	1.5	1.2
	2.9	3.8	2.8	2.9	2.7	2.7	2.2
Revenues/Receivables	39	46	42	38	43	44	48
	58	69	60	57	60	57	74
	77	88	75	82	86	75	113
Revenues/Working Capital	16.2	12.7	16.2	14.5	16.2	18.9	12.0
	9.1	12.1	10.3	7.6	8.6	9.2	6.9
	5.0	-98.5	6.0	4.7	2.3	5.0	-37.0
Revenues/Worth	11.7	11.8	12.0	10.4	8.8	10.5	6.8
	7.8	10.4	8.3	6.7	5.6	6.2	4.8
	4.3	5.9	5.4	4.1	3.6	4.8	2.4
% Profit Before Taxes/Worth	43.2	58.2	36.2	50.7	36.7	34.3	47.1
	17.6	37.0	17.2	17.5	21.8	23.2	18.7
	6.8	11.6	8.7	3.1	3.5	.0	3.5
% Profit Before Taxes/Tot. Assets	17.0	14.7	14.2	18.9	18.2	16.1	19.3
	7.4	9.1	7.4	6.9	8.6	12.4	5.4
	2.0	3.5	3.5	.6	2.7	3.8	1.5
Contract Revenues	$1941955M	$159514M	$286988M	$67036M	$128514M	$63866M	$9840M
Total Assets	608403M	64058M	109105M	26303M	116363M	24562M	5702M

SOURCE: Robert Morris Associates. See note 2 at end of chapter.

CONTRACTORS—ROOFING & SHEET METAL WORK

32 statements ended on or about June 30, 1972
47 statements ended on or about December 31, 1972

CONTRACT REVENUE SIZE Number of Statements	Under $1MM 43	$1MM & less than $10MM 32	$10MM & less than $50MM	$50MM & over	All sizes 79
ASSETS	%	%	%	%	%
Cash	10.3	9.4			10.8
A/R-Progress Billings	35.2	39.2			43.3
A/R-Current Retainage	2.6	6.1			5.6
Unbilled Work in Progress	6.9	7.6			5.2
Costs in Excess of Billings	2.3	.3			1.0
Other Current Assets	10.9	10.6			12.8
Total Current	68.0	73.2			78.7
Equipment	16.4	11.6			10.9
Real Estate	6.8	2.4			2.4
Joint Ventures	1.6	.6			.5
Other Non-Current	7.2	12.2			7.4
Total	100.0	100.0			100.0
LIABILITIES					
Notes Payable-Bank	6.4	10.0			8.4
Accounts Payable-Trade	14.7	20.0			18.3
Accounts Payable-Retainage	.0	.0			.0
Federal Income Tax Payable	2.4	3.1			5.0
Billings in Excess of Costs	.8	.1			4.5
Contract Advances	1.0	.2			.2
Prov. for Loss on Contr in Prg.	.0	3.3			3.2
Current Maturities LT Debt	2.2	1.7			2.3
Other Current Liabilities	12.9	14.4			10.5
Total Current	40.3	52.7			52.5
LT Liabilities, Unsub.	7.8	6.2			8.2
Total Unsubordinated Debt	48.1	58.9			60.7
Subordinated Debt	1.3	.1			.2
Tangible Net Worth	50.6	41.1			39.1
Total	100.0	100.0			100.0
INCOME DATA					
Contract Revenues	100.0	100.0			100.0
Costs of Work Performed	73.4	73.9			76.9

CONTRACTORS—STRUCTURAL STEEL ERECTION

18 statements ended on or about June 30, 1972
32 statements ended on or about December 31, 1972

CONTRACT REVENUE SIZE Number of Statements	Under $1MM 14	$1MM & less than $10MM 34	$10MM & less than $50MM	$50MM & over	All sizes 50
ASSETS	%	%	%	%	%
Cash	7.7	7.5			5.1
A/R-Progress Billings	44.9	38.0			26.7
A/R-Current Retainage	.8	8.3			9.2
Unbilled Work in Progress	1.1	6.2			15.3
Costs in Excess of Billings	2.4	1.2			.7
Other Current Assets	7.4	6.6			4.0
Total Current	64.2	67.9			61.1
Equipment	12.2	16.4			26.6
Real Estate	11.7	3.3			2.3
Joint Ventures		2.3			1.2
Other Non-Current	11.9	10.0			8.8
Total	100.0	100.0			100.0
LIABILITIES					
Notes Payable-Bank	7.1	10.2			7.7
Accounts Payable-Trade	18.2	13.2			13.4
Accounts Payable-Retainage	.0	1.0			.5
Federal Income Tax Payable	1.4	4.4			2.7
Billings in Excess of Costs	10.1	1.8			1.4
Contract Advances	.0	.2			.1
Prov. for Loss on Contr in Prg.	.0	.0			.0
Current Maturities LT Debt	2.7	4.0			2.2
Other Current Liabilities	5.5	8.8			15.1
Total Current	45.0	43.6			43.1
LT Liabilities, Unsub.	11.8	8.1			12.1
Total Unsubordinated Debt	56.8	51.6			55.2
Subordinated Debt	1.2	3.8			2.1
Tangible Net Worth	42.0	44.5			42.7
Total	100.0	100.0			100.0
INCOME DATA					
Contract Revenues	100.0	100.0			100.0
Costs of Work Performed	74.7	71.3			80.3

	Col 1	Col 2	Col 3		Col 4	Col 5	Col 6
Gross Profit	26.6	26.1	23.1		25.3	28.7	19.7
All Other Expense Net	24.1	22.0	19.3		23.8	23.3	19.3
Profit Before Taxes	2.5	4.1	3.9		1.5	5.4	.5
RATIOS							
Quick	1.8	1.7	1.7		2.0	1.9	2.0
	1.1	1.1	1.1		1.4	1.2	1.2
	.8	.9	.8		.9	.8	.8
Current	2.5	1.9	2.2		2.2	2.4	2.4
	1.6	1.5	1.5		1.8	1.4	1.5
	1.2	1.1	1.2		1.0	1.1	1.1
Fixed/Worth	.2	.2	.2		.2	.2	.2
	.4	.4	.4		.5	.5	.5
	.7	.6	.6		1.1	.9	1.0
Debt/Worth	.5	.8	.7		.5	.8	.7
	1.1	1.3	1.2		1.3	1.5	1.5
	2.5	2.8	2.9		6.0	2.9	2.9
Revenues/Receivables	**36** 10.0	**49** 7.3	**40** 9.0		**55** 6.5	**53** 6.8	**52** 6.9
	47 7.6	**64** 5.6	**58** 6.2		**71** 5.1	**77** 4.7	**75** 4.8
	71 5.1	**78** 4.6	**77** 4.7		**116** 3.1	**92** 3.9	**92** 3.9
Revenues/Working Capital	17.6	13.7	14.4		13.4	13.0	13.4
	7.6	9.6	8.9		7.0	7.6	7.2
	4.3	5.8	4.5		-31.6	3.4	3.4
Revenues/Worth	9.9	12.0	10.7		9.0	8.4	8.4
	5.0	6.5	5.8		3.6	5.3	4.9
	3.2	4.3	3.7		1.8	3.8	3.0
% Profit Before Taxes/Worth	33.5	44.0	39.0		45.4	42.8	43.0
	11.8	28.5	17.1		13.6	29.7	24.1
	2.8	10.8	6.1		2.5	17.7	10.1
% Profit Before Taxes/Tot. Assets	11.9	19.4	16.8		11.7	18.4	16.8
	7.1	8.0	7.6		5.0	10.1	8.9
	.0	3.5	1.6		-8.3	4.1	3.1
Contract Revenues	$22782M	$74501M	$160341M		$7619M	$102991M	$158156M
Total Assets	8603M	31004M	63817M		3762M	48370M	90756M

Source: Robert Morris Associates.   See note 2 at end of chapter.

161

CONTRACTORS—PLASTERING, DRYWALL, ACOUSTICAL & INSULATION WORK

30 statements ended on or about June 30, 1972
55 statements ended on or about December 31, 1972

CONTRACT REVENUE SIZE	Under $1MM	$1MM & less than $10MM	$10MM & less than $50MM	$50MM & over	All sizes
Number of Statements	37	44			85
ASSETS	%	%	%	%	%
Cash	8.5	6.9			6.0
A/R-Progress Billings	41.5	48.5			38.5
A/R-Current Retainage	6.6	5.5			5.8
Unbilled Work in Progress	8.9	7.6			9.5
Costs in Excess of Billings	2.0	2.9			1.5
Other Current Assets	7.5	8.7			9.7
Total Current	74.8	80.2			71.0
Equipment	13.2	11.6			7.5
Real Estate	4.5	1.7			7.7
Joint Ventures	.8	.1			.1
Other Non-Current	6.6	6.5			13.8
Total	100.0	100.0			100.0
LIABILITIES					
Notes Payable-Bank	10.0	10.2			11.3
Accounts Payable-Trade	17.5	20.2			16.3
Accounts Payable-Retainage	.6	.0			.0
Federal Income Tax Payable	2.1	3.2			3.0
Billings in Excess of Costs	.6	3.9			2.4
Contract Advances	.1	.6			.3
Prov. for Loss on Contr in Prg.	.3				.0
Current Maturities LT Debt	2.7	3.1			3.3
Other Current Liabilities	13.2	10.5			10.1
Total Current	47.0	51.5			46.9
LT Liabilities, Unsub.	12.7	9.4			16.0
Total Unsubordinated Debt	59.7	60.9			62.9
Subordinated Debt	1.4	5.0			2.4
Tangible Net Worth	38.9	34.1			34.6
Total	100.0	100.0			100.0
INCOME DATA					
Contract Revenues	100.0	100.0			100.0
Costs of Work Performed	73.2	81.9			79.6

CONTRACTORS—PLUMBING, HEATING (EXCEPT ELECTRIC), & AIR CONDITIONING

121 statements ended on or about June 30, 1972
214 statements ended on or about December 31, 1972

CONTRACT REVENUE SIZE	Under $1MM	$1MM & less than $10MM	$10MM & less than $50MM	$50MM & over	All sizes
Number of Statements	126	191	10		335
ASSETS	%	%	%	%	%
Cash	8.9	9.8	7.9		10.4
A/R-Progress Billings	37.2	40.7	39.3		43.5
A/R-Current Retainage	4.8	10.6	15.4		12.8
Unbilled Work in Progress	3.8	4.5	4.5		2.5
Costs in Excess of Billings	4.2	4.7	.8		4.2
Other Current Assets	10.2	9.3	14.0		9.8
Total Current	69.1	79.6	81.7		83.2
Equipment	14.7	10.5	4.5		8.6
Real Estate	9.1	2.8	.2		2.4
Joint Ventures	.0	.6	1.5		.6
Other Non-Current	7.1	6.5	12.1		5.2
Total	100.0	100.0	100.0		100.0
LIABILITIES					
Notes Payable-Bank	6.7	7.3	7.6		4.1
Accounts Payable-Trade	21.2	24.2	31.2		25.0
Accounts Payable-Retainage	.5	1.9	3.0		3.5
Federal Income Tax Payable	1.8	3.1	3.5		5.5
Billings in Excess of Costs	5.6	6.0	2.2		8.3
Contract Advances	.8	1.3	2.4		.8
Prov. for Loss on Contr in Prg.	.1	.0			.6
Current Maturities LT Debt	2.3	1.4	1.6		1.1
Other Current Liabilities	10.0	10.6	9.7		11.0
Total Current	49.0	56.0	61.3		59.4
LT Liabilities, Unsub.	8.1	7.3	13.1		7.0
Total Unsubordinated Debt	57.0	63.3	74.4		66.3
Subordinated Debt	1.1	1.2	.0		.6
Tangible Net Worth	41.9	35.5	25.6		33.1
Total	100.0	100.0	100.0		100.0
INCOME DATA					
Contract Revenues	100.0	100.0	100.0		100.0
Costs of Work Performed	74.0	83.5	89.1		86.4

	C1	C2	C3	C4	C5	C6	C7
Gross Profit	26.8	18.1	20.4	26.0	16.5	10.9	13.6
All Other Expense Net	24.5	15.2	17.0	23.5	13.6	14.3	11.5
Profit Before Taxes	2.3	3.0	3.4	2.5	2.9	2.7	2.7
RATIOS							
Quick	1.9	1.5	1.7	1.6	1.5	1.2	1.5
	1.3	1.3	1.2	1.1	1.1	1.0	1.1
	.8	1.0	.9	.8	.9	.7	.9
Current	2.4	1.9	2.0	2.1	1.9	1.5	2.0
	1.5	1.5	1.5	1.5	1.5	1.4	1.5
	1.1	1.3	1.2	1.1	1.2	1.1	1.2
Fixed/Worth	.1	.2	.1	.2	.1	.0	.2
	.2	.3	.3	.4	.3	.0	.3
	.6	.6	.6	.9	.6	.3	.7
Debt/Worth	.6	1.1	1.0	.5	.9	1.7	.8
	1.6	2.0	1.8	1.3	1.8	3.1	1.7
	2.6	2.8	2.7	2.7	3.3	4.1	3.2
Revenues/Receivables	50	47	49	42	47	48	45
	67	63	64	55	61	72	60
	80	84	84	80	77	95	78
Revenues/Working Capital	14.5	18.6	17.2	14.5	18.3	29.8	17.7
	7.2	10.5	9.5	8.2	10.1	14.0	10.1
	3.7	7.4	5.5	4.6	6.3	8.4	5.7
Revenues/Worth	11.0	13.0	11.5	8.4	12.6	15.7	11.5
	6.8	8.9	8.5	6.1	8.8	10.0	7.8
	3.9	6.6	5.3	4.0	5.6	6.5	4.9
% Profit Before Taxes/Worth	39.8	32.5	32.5	35.2	36.6	38.8	35.2
	16.6	21.4	20.6	18.1	17.2	29.9	18.7
	4.6	12.0	8.0	5.0	7.0	10.0	6.8
% Profit Before Taxes/Tot. Assets	14.7	12.0	12.8	13.9	12.9	11.5	12.9
	5.7	6.9	6.5	6.2	6.9	8.0	6.6
	2.4	3.7	3.2	1.7	2.0	2.9	2.1
Contract Revenues	$18837M	$116758M	$205430M	$66980M	$601532M	$165987M	$1529400M
Total Assets	7443M	44366M	94893M	27946M	211724M	61598M	534968M

Source: Robert Morris Associates. See note 2 at end of chapter.

163

CONTRACTORS—MASONRY, STONE SETTING & OTHER STONEWORK

16 statements ended on or about June 30, 1972
30 statements ended on or about December 31, 1972

CONTRACT REVENUE SIZE	Under $1MM	$1MM & less than $10MM	$10MM & less than $50MM	$50MM & over	All sizes
Number of Statements	20	25			46
ASSETS	%	%	%	%	%
Cash	8.5	10.9			15.0
A/R-Progress Billings	39.1	37.6			35.3
A/R-Current Retainage	6.4	11.2			14.5
Unbilled Work in Progress	15.3	11.2			9.3
Costs in Excess of Billings	2.2	3.5			2.7
Other Current Assets	1.4	5.4			4.5
Total Current	72.9	79.9			81.3
Equipment	19.0	11.9			10.1
Real Estate	2.6	3.6			4.2
Joint Ventures	.0	1.5			1.1
Other Non-Current	5.5	3.1			3.4
Total	100.0	100.0			100.0
LIABILITIES					
Notes Payable-Bank	14.3	15.3			12.0
Accounts Payable-Trade	15.1	18.8			15.8
Accounts Payable-Retainage	.6	.0			.0
Federal Income Tax Payable	.6	3.7			4.7
Billings in Excess of Costs	10.7	7.0			13.9
Contract Advances	.0	1.3			.9
Prov. for Loss on Contr in Prg.	.0	.0			.0
Current Maturities LT Debt	3.5	1.6			1.5
Other Current Liabilities	9.2	13.0			11.8
Total Current	53.3	60.9			60.5
LT Liabilities, Unsub.	4.2	5.5			4.3
Total Unsubordinated Debt	57.5	66.4			64.8
Subordinated Debt	1.3	1.8			1.4
Tangible Net Worth	41.2	31.8			33.8
Total	100.0	100.0			100.0
INCOME DATA					
Contract Revenues	100.0	100.0			100.0
Costs of Work Performed	77.8	83.8			82.5

CONTRACTORS—PAINTING, PAPER HANGING & DECORATING

10 statements ended on or about June 30, 1972
34 statements ended on or about December 31, 1972

CONTRACT REVENUE SIZE	Under $1MM	$1MM & less than $10MM	$10MM & less than $50MM	$50MM & over	All sizes
Number of Statements	25	19			44
ASSETS	%	%	%	%	%
Cash	13.5	9.4			10.3
A/R-Progress Billings	37.0	43.5			42.0
A/R-Current Retainage	4.2	2.9			3.2
Unbilled Work in Progress	6.9	8.9			8.5
Costs in Excess of Billings	.1	.4			.4
Other Current Assets	8.3	5.5			6.1
Total Current	70.0	70.6			70.5
Equipment	14.7	12.7			13.1
Real Estate	3.9	3.6			3.6
Joint Ventures	.0	1.1			.8
Other Non-Current	11.3	12.1			11.9
Total	100.0	100.0			100.0
LIABILITIES					
Notes Payable-Bank	13.5	12.5			12.7
Accounts Payable-Trade	12.8	13.6			13.5
Accounts Payable-Retainage	.1	.5			.4
Federal Income Tax Payable	1.8	2.7			2.5
Billings in Excess of Costs	2.6	2.0			2.1
Contract Advances	.0	5.2			4.1
Prov. for Loss on Contr in Prg.	.0	.0			.0
Current Maturities LT Debt	3.3	1.7			2.1
Other Current Liabilities	9.8	9.8			9.8
Total Current	44.0	48.1			47.2
LT Liabilities, Unsub.	8.7	5.3			6.1
Total Unsubordinated Debt	52.7	53.4			53.2
Subordinated Debt	3.6	1.9			2.2
Tangible Net Worth	43.7	44.8			44.5
Total	100.0	100.0			100.0
INCOME DATA					
Contract Revenues	100.0	100.0			100.0
Costs of Work Performed	73.8	80.0			78.6

Gross Profit	22.2	16.2	17.5	26.2	20.0	21.4
All Other Expense Net	18.8	12.6	12.4	21.3	16.4	17.5
Profit Before Taxes	3.5	3.5	5.1	4.9	3.6	3.9
RATIOS						
Quick	1.8 / 1.3 / 1.0	1.7 / 1.2 / .9	1.7 / 1.3 / .9	1.9 / 1.4 / .6	1.5 / 1.2 / .9	1.9 / 1.2 / .8
Current	2.1 / 1.5 / 1.2	1.8 / 1.4 / 1.1	2.0 / 1.5 / 1.1	2.2 / 1.6 / 1.1	1.8 / 1.5 / 1.3	2.2 / 1.6 / 1.2
Fixed/Worth	.4 / .6 / 1.0	.2 / .4 / .9	.2 / .5 / .9	.2 / .4 / .8	.1 / .3 / .5	.2 / .4 / .7
Debt/Worth	.6 / 1.2 / 3.2	1.1 / 1.6 / 3.4	.9 / 1.4 / 3.4	.5 / 1.1 / 3.9	.7 / 1.1 / 2.6	.6 / 1.2 / 3.5
Revenues/Receivables	**42** 8.6 / **61** 5.9 / **75** 4.8	**51** 7.0 / **63** 5.7 / **84** 4.3	**43** 8.3 / **61** 5.9 / **82** 4.4	**55** 6.6 / **63** 5.7 / **77** 4.7	**51** 7.1 / **58** 6.2 / **90** 4.0	**51** 7.0 / **59** 6.1 / **77** 4.7
Revenues/Working Capital	18.0 / 9.1 / 5.0	19.0 / 8.8 / 5.7	19.0 / 8.9 / 5.3	10.5 / 6.2 / 3.6	18.5 / 10.3 / 5.9	16.6 / 8.1 / 4.4
Revenues/Worth	12.6 / 7.2 / 4.1	13.3 / 7.6 / 5.5	13.0 / 7.5 / 4.2	7.5 / 5.3 / 2.4	12.4 / 7.0 / 3.8	12.4 / 6.1 / 3.2
% Profit Before Taxes/Worth	57.5 / 33.4 / 6.4	52.1 / 22.6 / 8.3	59.3 / 28.1 / 6.9	60.6 / 32.9 / 12.1	42.6 / 18.0 / 11.6	56.6 / 25.7 / 13.1
% Profit Before Taxes/Tot. Assets	20.0 / 11.3 / 4.6	17.0 / 7.3 / 3.3	19.7 / 10.2 / 3.7	24.8 / 11.7 / 1.6	13.6 / 8.0 / 5.7	19.8 / 10.1 / 4.9
Contract Revenues	$8758M	$61141M	$84480M	$10492M	$38017M	$48590M
Total Assets	3085M	21824M	31544M	4002M	14158M	18160M

Source: Robert Morris Associates. See note 2 at end of chapter.

CONTRACTORS—EXCAVATING & FOUNDATION WORK

30 statements ended on or about June 30, 1972
49 statements ended on or about December 31, 1972

CONTRACT REVENUE SIZE	Under $1MM	$1MM & less than $10MM	$10MM & less than $50MM	$50MM & over	All sizes
Number of Statements	33	40			79
	%	%	%	%	%
ASSETS					
Cash	6.5	8.6			11.3
A/R-Progress Billings	28.7	23.0			23.0
A/R-Current Retainage	5.4	8.9			11.0
Unbilled Work in Progress	3.2	3.2			2.7
Costs in Excess of Billings	.2	1.0			.7
Other Current Assets	2.6	6.2			7.8
Total Current	46.7	50.9			56.6
Equipment	43.1	32.6			30.7
Real Estate	5.6	2.7			2.3
Joint Ventures	.0	2.5			2.0
Other Non-Current	4.7	11.4			8.4
Total	100.0	100.0			100.0
LIABILITIES					
Notes Payable-Bank	9.6	5.7			4.3
Accounts Payable-Trade	14.6	18.1			16.6
Accounts Payable-Retainage	.8	.6			1.3
Federal Income Tax Payable	.8	2.3			6.6
Billings in Excess of Costs	.1	1.4			1.9
Contract Advances	1.9	.3			.3
Prov. for Loss on Contr in Prg.	.0	.0			.0
Current Maturities LT Debt	6.0	7.2			5.9
Other Current Liabilities	6.7	6.3			6.6
Total Current	40.5	41.8			43.5
LT Liabilities, Unsub.	17.5	16.7			15.5
Total Unsubordinated Debt	58.0	58.5			59.1
Subordinated Debt	4.3	.4			.6
Tangible Net Worth	37.8	41.1			40.4
Total	100.0	100.0			100.0
INCOME DATA					
Contract Revenues	100.0	100.0			100.0
Costs of Work Performed	68.9	63.4			72.2

CONTRACTORS—TERRAZZO, TILE, MARBLE & MOSAIC WORK

6 statements ended on or about June 30, 1972
13 statements ended on or about December 31, 1972

CONTRACT REVENUE SIZE	Under $1MM	$1MM & less than $10MM	$10MM & less than $50MM	$50MM & over	All sizes
Number of Statements	15				19
	%	%	%	%	%
ASSETS					
Cash	7.9				6.7
A/R-Progress Billings	31.1				37.3
A/R-Current Retainage	4.2				2.0
Unbilled Work in Progress	17.9				20.5
Costs in Excess of Billings	.4				.2
Other Current Assets	11.4				11.9
Total Current	72.9				78.5
Equipment	6.3				5.5
Real Estate	12.6				7.3
Joint Ventures	.6				.3
Other Non-Current	7.6				8.4
Total	100.0				100.0
LIABILITIES					
Notes Payable-Bank	10.7				13.9
Accounts Payable-Trade	10.6				15.2
Accounts Payable-Retainage	.1				.0
Federal Income Tax Payable	1.2				1.2
Billings in Excess of Costs	1.8				.9
Contract Advances	.0				
Prov. for Loss on Contr in Prg.	.0				7.7
Current Maturities LT Debt	4.1				3.2
Other Current Liabilities	12.3				14.4
Total Current	40.8				56.4
LT Liabilities, Unsub.	18.8				11.5
Total Unsubordinated Debt	59.5				67.9
Subordinated Debt	.8				.4
Tangible Net Worth	39.7				31.7
Total	100.0				100.0
INCOME DATA					
Contract Revenues	100.0				100.0
Costs of Work Performed	73.6				77.9

Gross Profit	31.1	36.6	27.8	26.4	22.1
All Other Expense Net	24.0	30.9	20.2	22.7	19.1
Profit Before Taxes	7.1	5.7	7.8	3.7	3.0
RATIOS					
Quick	1.3	1.6	1.4	3.4	2.7
	.8	1.0	1.0	1.7	1.9
	.7	.8	.8	.7	.6
Current	1.6	1.9	1.7	4.2	3.4
	1.0	1.3	1.2	2.1	1.8
	.8	1.0	.9	1.6	1.3
Fixed/Worth	.8	.4	.5	.1	.1
	1.4	1.0	1.0	.2	.2
	2.2	1.6	1.7	.9	.6
Debt/Worth	.9	.7	.8	.3	.3
	1.7	1.4	1.5	.7	.8
	2.8	2.8	2.8	1.3	2.7
Revenues/Receivables	45	52	51	44	44
	66	66	66	55	59
	86	78	78	97	97
	8.0	6.9	7.0	8.1	8.1
	5.5	5.5	5.5	6.6	6.1
	4.2	4.6	4.6	3.7	3.7
Revenues/Working Capital	10.1	19.0	15.9	8.7	9.6
	3.8	7.5	6.6	5.5	5.5
	-15.6	-84.2	-38.2	2.1	2.1
Revenues/Worth	8.0	8.1	8.2	7.4	7.9
	4.2	5.6	5.4	5.2	5.7
	2.9	3.0	3.0	2.3	2.5
% Profit Before Taxes/Worth	59.3	45.0	56.6	29.7	29.7
	23.3	23.3	23.3	21.9	22.0
	6.9	12.9	9.5	7.0	8.7
% Profit Before Taxes/Tot. Assets	16.3	15.2	17.2	13.3	12.2
	9.0	9.7	9.7	6.3	5.6
	2.5	5.7	3.6	1.9	2.4
Contract Revenues	**$18592M**	**$98429M**	**$220836M**	**$7772M**	**$17195M**
Total Assets	**10602M**	**54886M**	**114930M**	**3857M**	**8242M**

SOURCE: Robert Morris Associates.   See note 2 at end of chapter.

167

would show the estimated current gross profit compared to that at the time of bidding.

When major subcontracts are being negotiated, it is customary to request a current financial statement from the proposed subcontractor to submit with the purchasing man's recommendation for approval by the contractor or the financial officer of the company. Practices differ, but a file of such statements is of value for comparison's sake as later subcontracts are awarded.

Financial statements of other subcontractor bidders submitted for information during negotiations should be filed alphabetically with other financial data and résumés.

There have always been cycles in the building business when construction slows down, money is hard to come by, and contractors and subcontractors fail. Going back 50 years or so, after World War I funds became reasonably adequate, and contractors were able to pursue their work with less use of credit than became the custom later. Because the jobs were more nearly standard and less complex, their completion dates could be forecast more easily. There was no air conditioning and no special electronics equipment which now have primary and secondary effects over most trades on a building contract.

As Robert G. Cerny points out,

> This era came to a crashing halt in the great depression. Construction stopped, credit dried up, and the industry faced a grim battle for survival. There developed an industry wide conspiracy to operate with little capital and less credit. Sureties issued bonds based upon faith and performance. Suppliers and subcontractors issued lien waivers on unpaid bills to the general contractor. The waivers were certified by the architect and submitted to the client-owner for payment. The payment was eagerly divided among the creditors. Everyone cooperated and the conspiracy worked.[5]

Unfortunately the emergency patterns developed during the Depression still persist and are often exploited by the unscrupulous. Speculative building in recent years has aggravated the tendency to operate "off the cuff."

One federal judge recently described the construction industry as a jungle of confusion, stating,

> It is the only contemporary industry in which millions of dollars of the client's money is paid out in the pious hope that it will ultimately be paid to the subcontractors, suppliers and other contributing components.[6]

For the general contractor, therefore, a primary objective should be to try to avoid involvement in financial problems with subcontractors.

Failing that, however, when there is involvement, it is necessary to determine ways and means to handle the resulting situations.

Noninvolvement is easier said than done. There will always be risks because of the many variables which contribute to the picture. The proper selection of bidders, with awards only to qualified subcontractors, is made most frequently with experience. Just as one poor automobile driver can cause an accident despite the other driver's precautions, the variables can greatly affect the commitments.

One must face the fact that if every detail in the agreement is "safe," the price may be too high. There will be no contract and ultimately no further work for the contractor. Actually, while low bidders as a class are usually the first to go bankrupt, they are followed rather closely by the high bidders who are unable to secure enough jobs.

MORE CHECKPOINTS WHEN CONTRACTING

Careful consideration of the breakdown of the subcontract price is advisable when it is first submitted. It should not be weighted heavily toward initial payments, for a 10 percent overage would reduce by the same amount the funds available for completing the last 10 percent of the job.

Some financial problems can be foreseen, and efforts should be made to resolve them before they occur. For instance, one of the most common problems concerns arrangements to authorize changes. Outright additions usually can be, and are, priced promptly and a decision secured. When the change is complicated by omissions, revisions, additions, and different operating procedures, discussion may be necessary after the estimate has been submitted before it can be authorized or rejected. Any undue period of uncertainty or waiting handicaps the subcontractor and affects job progress. All parties have a responsibility and an interest to process these estimates promptly. A knowledge of the subcontractor and an "eyeball to eyeball" conference can go a long way toward securing cooperation in handling estimate changes. The competent subcontractor knows from experience how necessary it is to have his requisitions paid when due. In the negotiations, if greater attention is given to assurances of prompt submission of change estimates, a good start will have been made.

SUBCONTRACT PAYMENTS

Another way to prevent financial repercussions is for the contractor or his superintendent to realistically approve the value of work completed and in place. There are risks involved in making payments to subcontractors covering only materials delivered to the site. Waiting for the material to be installed allows a period of time to flush out delinquencies and nonpayment to vendors for the materials.

Some owners are willing to permit payment for delivered materials, on the basis of which title passes to the owner, and the value of the material may be properly listed on the monthly requisition.

Among the good reasons which furnish grounds for material payments would be:

1. The material was delivered according to an approved schedule and, because of a slippage in the field work, will not be needed for a month or two, perhaps longer.

2. The materials are in short supply, and while they still can be obtained it is desirable to receive on-the-job deliveries.

3. The subcontractor is pressed for funds, but to pay him could be risky. If it is new material coming from the subcontractor's stock that has been paid for, it is justifiable; but the request for payment represents a warning signal that finances are tight. Naturally there is need to verify that the vendors actually have been paid.

4. There is no question of the subcontractor's unloading an oversupply of pipe, for instance, which will clutter up the job, require rehandling, and generally inconvenience other subcontractors.

5. The subcontractor will absorb all rehandling and storage charges if off the site. Usually owner's insurance coverage may apply only to on-site storage. This should be verified and, if necessary, the subcontractor should agree to assume premium costs.

6. The architect and owner approve.

If, to keep the job moving, some payment seems advisable and both the architect and owner approve, and if it is a bonded job and the bonding company's agreement is secured, payment may be advanced. The contractor in such cases would request the owner's consent to include the advance in the next monthly requisition.

Where money is in short supply, the situation can become rugged for both the subcontractor and the contractor who must bear the ultimate responsibility for financing the work to keep it moving. Advancing money to a subcontractor is a surefire means of getting action on the job. It can also be extremely risky unless adequate safeguards are given that past-due accounts on the particular job are being paid and means agreed upon for settlement.

ANOTHER RED FLAG

It might be noted that management changes in many subcontractors' organizations; occasions arise when firms with outstanding reputations are sold to speculators whose sole purpose, it develops later, was to secure the assets of the firm and depart.

The contractor should not be lulled into an acceptance of business relationships without evidence that the new management is operating in a trustworthy manner. There are times when credit reports lag and unexpectedly indicate defaults rather than a noticeable tightening of a subcontractor's financial position.

BILL OF SALE

It should be recognized that advancing payments for materials usually represents a change in the subcontract terms that should not be made casually. If the amount involved is considerable, a properly executed bill of sale (Appendix C, Form 3) would legally transfer title and would avoid repercussions later if the subcontractor fails to pay his vendor.

AFFIDAVITS

Many contractors use payment affidavits (Appendix C, Form 4) which establish a good legal position with respect to the statement of a subcontractor's accounts. While it does not follow that all subcontractors give these documents the consideration that they deserve, a contractor will find them comforting to have if a default occurs and the accuracy of the affidavit comes into question. Affidavits commonly are used to obtain a statement of the subcontractor's accounts when a change is being considered with respect to the subcontract reserve withholding.

LIENS

State laws vary greatly with respect to liens which may be filed for non-payment of claims. In some states an owner may be responsible for paying a second time if reimbursement has not been properly made. For instances of these differences, the reader is referred to examples applying in the following states: California, Connecticut, Florida, Illinois, Massachusetts, Michigan, Minnesota, New Jersey, New York, Ohio, and Pennsylvania.[7] Where laws are less stringent, the tendency may be to lessen the protection for a vendor or banker and consequently increase the difficulty of securing credit.[8]

Many financial institutions decline to make loans that are junior in claim position to the other seven creditors given in order as:

1. The owner
2. All material men and suppliers

3. All labor
4. All subcontractors and all subs of subs
5. All material men and suppliers of subcontractors
6. All federal and state taxes
7. The bonding companies

There are also variations involving notification in connection with unpaid accounts. Where the contractor has been made aware of such delays, it is good judgment to take steps to see that the parties have come to an understanding about payment before releasing the current requisition payment to the delinquent subcontractor.

PROS AND CONS OF UNIT PRICING

Unit pricing is sometimes a dubious tool for pricing work because many changes cannot be fairly evaluated by this method. Despite its shortcomings, however, any precontract understanding regarding processing of extras is useful and will expedite approvals once the changes start. There is a mutual benefit in facilitating all change-order work that pays off in avoiding delays in the field.

If unit prices are established which recognize the time factor of when the change would apply, one source of later discussion has been obviated. For instance, on some jobs an agreement is reached that changes on structural steel made prior to the preparation of shop drawings or the placing of the mill order will be at lower rates than if made later. Similarly, increase in the weight of individual members might be priced to provide only for the increased material cost. In most cases, erection is based on a price per piece, and the increased weight is not of significance to the erector.

When unit prices are to be used, they should be checked for the reasonableness at the time the contract is negotiated. In the same way, contract breakdowns should not be casually approved but should be considered sufficiently prior to acceptance to avoid unwelcome retroactive corrections.

RESERVE WITHHOLDING

Most construction agreements specify that there will be a 10 percent withholding until such time as the work is completed and approved. Final payment frequently is specified to be made within 30 days after such approval.

On a job of considerable duration and of some magnitude, it may be agreed to make no further withholdings after the contract is 50 percent complete, thereby resulting in a 5 percent reserve at the completion of the job.

On some jobs, considerable work may have been performed on changes for which an exact amount may not have been agreed upon. To avoid confusion, each change should be given an estimate number. While advance payments on account can be authorized for work performed in such cases, with the usual 10 percent reserve, eventually the audited and approved change estimate would become part of the contract, and the advance payments should be deducted from the final approved estimate to keep the account clear. When there is a large volume of change estimates, there may be difficulty in determining to which estimate the advance applies.

In some cases, a provisional approval can be given to a group of change estimates, using a 20 percent reserve withholding, until such time as the estimates have been finally approved. Then correcting change-order additions or deductions (Figs. 16 and 17) are used to reflect the final amounts.

This method permits partial payment and identification of the changes for subcontractors and their own subcontractors, which would not be the case if the advance is not proportionately allocated to specific changes.

BONDS

Generally speaking, there are three types of bonds used in the construction business. The *bid bond*, for public work, usually in the amount of 5 percent or 10 percent of the proposal figure, is an agreement to accept a contract at the stated price of the bid. There needs to be a clear understanding of the purpose by the surety and by the concern being bonded. The bond is for one purpose only, and the bonding limit is inapplicable if the contract is not secured.

The *performance bond* (Fig. 14) is issued as a guarantee by the surety company that the contract will be performed by the subcontractor and, in the event of a failure, the surety will complete the work.

The *payment bond* (Fig. 15) is an agreement to guarantee payment of all proper claims and accounts on the project work.

For performance and payment bonds the surety company issuing the bond agrees to guarantee performance and/or payment. If the subcontractor performs properly, the obligation has been satisfied. It is only when there is a failure that the surety is called upon to perform. In most cases, the surety will have priority over loans by a bank, but the situations are such that joint action by both bank and surety may extend help to the subcontractor rather than permit a default to occur.

PERFORMANCE BOND

KNOW ALL MEN BY THESE PRESENTS, That ....,
<small>(name of Subcontractor)</small>

a .... corporation, of ....,
<small>(address)</small>

as Principal (hereinafter called the "Principal"), and ....,
<small>(name of Surety)</small>

a .... corporation, of ....,
<small>(address)</small>

as Surety (hereinafter called the "Surety"), are held and firmly bound unto A. B. C. CONSTRUCTION COMPANY, a Delaware corporation (hereinafter called the "Obligee"), in the sum of..............................
.... Dollars ($.......................), for the payment of which sum well and truly to be made, the Principal and Surety bind themselves, and their respective heirs, administrators, executors, successors and assigns, jointly and severally, firmly by these presents.

WHEREAS, the Principal and the Obligee have entered into a written agreement dated ....
.................., (hereinafter called the "Subcontract"), for the performance by the Principal, as Subcontractor, of ...., all as more fully described and mentioned in
<small>(description of the Subcontract work)</small>
said Subcontract, which Subcontract is hereby incorporated in and made a part of this Bond with the same force and effect as if fully set forth at length herein,

NOW, THEREFORE, THE CONDITION OF THIS OBLIGATION IS SUCH, that if the above bounden Principal shall well and truly perform all of the undertakings, covenants, terms, conditions and agreements of said Subcontract within the time provided therein and any extensions thereof that may be granted by the Obligee, and during the life of any guarantees contained in or required under said Subcontract, and shall also well and truly perform all the undertakings, covenants, terms, conditions and agreements of any and all modifications of said Subcontract that may hereafter be made, and shall indemnify and save harmless the Obligee of and from any and all loss, damage and expense, including costs and attorneys' fees which the Obligee may sustain by reason of the Principal's failure, neglect and/or refusal so to do, then this obligation shall be null and void; otherwise it shall remain in full force and effect.

The Surety, for value received, agrees that its obligation shall in nowise be impaired or affected by any change, extension of time, alteration, addition, omission or other modification in or of the said Subcontract or the work to be performed thereunder and does hereby waive notice of any and all such changes, extensions of time, alterations, additions, omissions, and/or other modifications.

The Surety, for value received, agrees, if requested to do so by the Obligee, to perform fully and complete the work mentioned and described in said Subcontract and any and all modifications thereof pursuant to and in accordance with the undertakings, covenants, terms, conditions and agreements thereof, if the Principal fails, neglects and/or refuses to so perform fully and complete said work, and further agrees to commence the performance and completion of said work within ten (10) days after notice thereof from the Obligee of such failure, neglect and/or refusal of the Principal and to perform and complete the same within the time required under said Subcontract and any and all modifications thereof as extended by the period of time elapsing between the date of such failure, neglect and/or refusal of the Principal and the date of the giving of such notice by the Obligee to the Surety.

This Bond shall be for the sole benefit of the Obligee, its successors and assigns.

IN WITNESS WHEREOF, the Principal and Surety have hereunto affixed their corporate seals and caused this Bond to be duly executed and acknowledged by their duly authorized officers or representatives this ....
day of ...., 19....

(Impress Corporate Seal)　　　　　　　.., Principal

　　　　　　　　　　　　　　　　By ....
<small>(Title)</small>

(Impress Corporate Seal)　　　　　　　.., Surety

　　　　　　　　　　　　　　　　By ....
<small>(Title)</small>

[Attach corporate acknowledgments of Principal and Surety]

FIG. 14   Performance Bond.

PAYMENT BOND

KNOW ALL MEN BY THESE PRESENTS, That ....,

(name of Subcontractor)

a .... corporation, of ....,

(address)

as Principal (hereinafter called the "Principal"), and ....,

(name of Surety)

a .... corporation, of ....,

(address)

as Surety (hereinafter called the "Surety"), are held and firmly bound unto A. B. C. CONSTRUCTION COMPANY, a Delaware corporation (hereinafter called the "Obligee"), in the sum of............................. .... Dollars ($........................), for the payment of which sum well and truly to be made, the Principal and Surety bind themselves, and their respective heirs, administrators, executors, successors and assigns, jointly and severally, firmly by these presents.

WHEREAS, the Principal and the Obligee have entered into a written agreement dated ...., (hereinafter called the "Subcontract"), for the performance by the Principal, as Subcontractor, of ...., all as more fully described and mentioned in

(description of the Subcontract work)
said Subcontract, which Subcontract is hereby incorporated in and made a part of this Bond with the same force and effect as if fully set forth at length herein,

NOW, THEREFORE, THE CONDITION OF THIS OBLIGATION IS SUCH, that if the above bounden Principal shall promptly pay all persons having just claims for (a) labor, materials, services, insurance, supplies, machinery, equipment, rentals, fuels, oils, implements, tools and/or appliances and any other items of whatever nature, furnished for, used or consumed in the prosecution of the work called for by said Subcontract and any and all modifications thereof, whether lienable or nonlienable and whether or not permanently incorporated in said work, (b) pension, welfare, vacation and/or other supplemental employee benefit contributions payable under collective bargaining agreements with respect to persons employed upon said work and (c) federal, state and local taxes and/or contributions required by law to be withheld and/or paid with respect to the employment of persons upon said work, then this obligation shall be null and void; otherwise it shall remain in full force and effect.

The Surety, for value received, agrees that no change, extension of time, alteration, addition, omission or other modification of said Subcontract or the work to be performed thereunder shall in any wise impair or affect its obligation on this Bond and does hereby waive notice of any and all such changes, extensions of time, alterations, additions, omissions and/or other modifications.

The Principal and the Surety, for value received, agree that this Bond shall inure to the benefit of all persons having just claims as aforesaid, whether or not they have any direct contractual relationship with the Principal, as well as to the benefit of the Obligee, and that such persons may maintain independent actions upon this Bond in their own names.

IN WITNESS WHEREOF, the Principal and Surety have hereunto affixed their corporate seals and caused this Bond to be duly executed and acknowledged by their duly authorized officers or representatives this .... day of ...., 19....

(Impress Corporate Seal)   ...., Principal

By ....

(Title)

(Impress Corporate Seal)   ...., Surety

By ....

(Title)

[Attach corporate acknowledgments for Principal and Surety]

FIG. 15   Payment Bond.

FINANCING ARRANGEMENTS

Financing a building project properly often becomes a problem for two rather fundamental reasons. Probably the most usual is that the general contractor and many of his subcontractors are not possessed of enough ready funds to cover the unusual expenses that often arise on building work. Another cause is that changes authorized while the project is underway not only may hold back progress, making the cost of performance greater than expected, but also may delay payments until the cost of such changes has been agreed upon.

Under normal conditions, the architect and the owner could expect the contractor to establish a realistic value for his work and to submit monthly requisitions on the 25th of each month; having received payment by the 10th of the following month, he will pay all his subcontractors by the 15th of the same month.

In practice, unfortunately, it seldom works so smoothly. The subcontractor is faced not only with meeting his weekly payrolls for the contract work, but also with meeting those expenses which have been incurred on changes or work in progress. Perhaps the owner delays the payment to the contractor for a few days, and a weekend intervenes; our subcontractor then must secure credit to be able to discount invoices for material or to pay income taxes or insurance premiums.

It is probably safe to say that material suppliers grant more credit to contractors than do the banks. The credit extensions, of course, are part of their sales programs. This practice has its dangers, because often the call for payment by many of the suppliers comes (when the gossip of a bid job becomes current) at the same time, which is generally the most inopportune.[9]

The broker or brokers who may be willing to furnish a loan to the owner are reluctant to take on the added risk of further loans. The subcontractor must find another source for his venture capital to keep the job moving.

Bankers are apt to feel that the financial statements of most subcontractors are relatively weak. It is only by developing a close working relationship—month by month—with the bank that the subcontractor can impress upon his banking contact that he knows his business and estimates consistently, thus warranting the extension of further credit.

Naturally, these possibilities must be explored in advance; otherwise the job grinds to a stop as soon as payments are not forthcoming when due. Larger companies with adequate capital can withstand the occasional accumulation of minor discrepancies, can use their own funds if critical situations develop, and can arrange for proper handling in the future.

Even so, the subcontractor's problems in securing additional financing, when unexpected emergencies and delays require it, may be of less significance than those of the owner investor whose involvement has been based on expected occupancy or income and who may be faced with construction loans for an extended period. This could force some investors out of the project and imperil the whole job as well as the contractor and subcontractors. Teamwork to prevent delays should be one of the objectives of contractor, subcontractors, owner, and architect.

FINANCIAL STATEMENTS

The American Institute of Certified Public Accountants has made a start in establishing businesslike bases for construction contractors' financial statements.[10] While many subcontractors will use their own style, which may not be as clear or as consistent as one might wish, there are some standards that the contractor should be familiar with.[11]

In studying a statement, for instance, some of the points to note would be:

1. How much working capital is available?
2. Are other sources of funds available?
3. Has the statement been prepared within the past year and preferably within the past 6 months?
4. Does the subcontractor discount his bills?
5. Is equipment paid for or covered by a mortgage? How much is due and to whom?
6. What is status of accounts payable regarding currency of invoices?
7. What is status of reserve withheld for taxes and fringe benefit payments? Union benefit payments?
8. What is status and cash value of life insurance for principals?
9. What represents realistic appraisal of amount of contracts completed now underway?
10. What is status of change estimates, i.e., amounts approved or in dispute?

WORKING CAPITAL

This represents the difference between the current assets and the current liabilities; bankers are normally willing to extend loans to their qualified contractor customers up to the amount of the working capital.

The problem for anyone analyzing financial statements frequently is to determine accurately just what constitutes the current assets. There are twilight zones where it is difficult to classify those assets which are obviously current—such as cash—and those that are deferred; for this reason some larger contractors omit the familiar term "current assets." Form lumber or material on the site ready to be incorporated into the work may properly be called a current asset, but the same materials piled up in the contractor's yard in the hope that maybe someday a use will be found for them should hardly be so considered.

Current liabilities are more easily listed, but here an absence of reserves to cover accrued taxes or collections which are to be paid over to welfare funds should raise questions regarding the omissions.

The currency of the accounts payable may not be shown on statements but is valuable information in analyzing the subcontractor's ability to discount invoices. Actually, one of the telltale indications of a subcontractor's financial position is his ability to take advantage of available cash and trade discounts when he is in a healthy financial condition.

If it were possible to forecast potential financing problems, they could be analyzed and provided for in estimating. Instead, the problems come unexpectedly from many quarters, and the contractor may need other sources for funds. A knowledgeable banker, who has confidence in the integrity and ability of his customer and who stands ready to help with reasonably priced loans, is one possibility. If this type of credit is not available, it can be very expensive for the subcontractor faced with exorbitant loan rates to provide funds to keep the job moving.

LIFE INSURANCE

Life insurance for the principals provides a measure of protection in bringing cash assets into the company at a time when they are most needed and could prevent forced sales and dissolution of potentially sound assets. In a somewhat similar category, there is need for well-thought-out plans for continuation and management of the company so that it will not be a subject of litigation on the part of the surviving controlling stockholders, to the detriment of the work under contract.

Reserves on contracts in progress or collateral offered by a subcontractor sometimes do not give the protection expected, since they may prove of no material value. Similarly, heavy mortgage payments on equipment may be manageable if the equipment can be kept operating; otherwise they can become an albatross that encourages a contractor to take work too cheaply.

From the contractor's or owner's point of view, there will always be a desire to secure competent subcontractors who can perform without credit problems. A good reputation in the community is one of the first requirements before any banker is likely to become interested in extending credit.

CREDIT FOR SPECIALTY SUBCONTRACTORS

For the newly organized subcontractor, one possibility involves an arrangement, such as proposed by the United California Bank, whereby credit is extended in certain cases to specialty subcontractors. Owner's payments and all the job's disbursements are routed through the bank.[12]

From the owner's or contractor's viewpoint, there is assurance of proper settlement of the accounts; but the subcontractor would not be given credit unless he met the bank's standards. In a sense, he can expect "big brother" assistance and support from the bank because the bank is satisfied with the validity of his contract and his expected profits. Although, eventually, this type of assistance may be more than the subcontractor desires, initially it gives an owner a sense of security about performance which he might not have with a less-well-financed operation. It gives the subcontractor a chance to prove his ability which, while it may be obvious, has never been tested in a "trial by fire."

NOTES—Chapter 7

1. James A. Bourke, "Lending to Contractors," address at National Electrical Contractors Association meeting, Atlanta, Ga.
2. ROBERT MORRIS ASSOCIATES—DISCLAIMER STATEMENT: "RMA cannot emphasize too strongly that their composite figures for each industry may *not* be representative of that entire industry (except by coincidence), for the following reasons:

    (1.) The only companies with a chance of being included in their study in the first place are those for whom their submitting banks have recent figures.

    (2) Even from this restricted group of potentially includable companies, those which are chosen, and the total number chosen, are not determined in any random or otherwise statistically reliable manner.

    (3) Many companies in their study have *varied* product lines; they are 'mini-conglomerates,' if you will. All they can do in these cases is categorize them by their *primary* product line, and be willing to tolerate any 'impurity' thereby introduced.

    In a word, don't automatically consider their figures as representative norms and don't attach any more or less significance to them than is indicated by the unique aspects of the data collection."
3. Robert Morris Associates, "Statement Studies," Philadelphia, Pa., 1973.
4. Further elaboration of these guidelines appears in W. R. Park, *The Strategy of Contracting for Profit*, Prentice-Hall, Englewood Cliffs, N.J., 1966.
5. Robert G. Cerny, "Construction Industry Finances—Musical Chairs with Money," Robert Morris Associates Conference, San Juan, P. R., Oct. 29, 1969.
6. Ibid.
7. N. Walker and T. K. Rohdenburg, *Legal Pitfalls in Architecture, Engineering, and Building Construction*, McGraw-Hill, New York, 1968, pp. 163–191.
8. James A. Bourke, "Financing Problems in the Construction Industry," annual meeting, National Electrical Contractors Associations, Atlanta, Ga., June 15, 1971.
9. Ibid.

10. The bulletin, or *Industry Audit Guide,* prepared by the committee on Contractor Accounting and Auditing and the Committee on Cooperation with Surety Companies of the American Institute of Certified Public Accountants, Inc., New York, lists sample comparative balance sheets and statements of income with explanatory notes, furnishing much information of value in analyzing statements.

11. The statement should clearly state which of the two generally accepted methods is being followed, namely: (*a*) percentages of completion method—which is preferable when estimates of costs to complete and extent of progress toward completion of long-term contracts are reasonably complete; (*b*) completed contract method—which is preferable when lack of dependable estimates or inherent hazards cause forecasts to be doubtful.

12. Frank E. Bristow (United California Bank, Los Angeles), "Specialty Contractors' Financing Program," Robert Morris Associates, Philadelphia.

CHAPTER EIGHT

Legal Problems

Over the years contractors, as a class, have been more concerned and interested in doing work than in being mindful of the legal implications of the contracts. They are primarily builders, not lawyers, and, unless they are alert, can be hurt by unnoticed onerous or unfair clauses. Just as competent financial advice is essential for the general contractor, legal counsel can make the business risks less hazardous and help out in situations that crop up from time to time.

The legal aspects of purchasing cover a wide range of situations beyond those having to do with buying for construction and subcontracting.

The basic definitions that help to categorize the laws relating to purchasing are well illustrated in the *Purchasing Handbook,* by George W. Aljian, which is a very instructive guide for both buyers and sellers.

The old axiom that "A little learning is a dangerous thing" applies especially to the legal problems that arise in bidding and in the execution and performance of contracts in the building-construction industry. There have been countless cases where slight variations in the wording of contracts or agreements have led to different decisions, and so it is difficult to come up with absolute guidelines. The cases reported in this chapter have not been chosen to demonstrate the correctness of the decisions, but rather are intended to point out the different results as they affect both contractors and subcontractors under a slightly different set of facts.

In the building business the cases frequently involve, among other questions, contract interpretations, errors in bidding, breach of contract, and damages. Often several different questions are involved in the same case.

The basic legal knowledge needed by the purchasing man, which has been termed "preventive law," may be compared to that which the motorist needs to keep his automobile in good running order. The wise motorist practices preventive maintenance in keeping the tires properly inflated, replacing oil filters, lubricating the bearings, and so forth. This does not mean that he must be a mechanic to drive the car. The more he knows, however, the better will be the results in operating. In limiting his application of legal principles to this preventive law, the purchasing man will also recognize the problems and situations which should be referred to professional counsel.[1,*]

Misunderstandings come about easily in the building business and take all manner of forms. One of the author's first experiences was with a sprinkler subcontractor in Chicago more than 40 years ago. He was a fine gentleman and universally respected.

Our discussion led to an "agreement," the final terms of which he confirmed as he was leaving the office. The following day the written confirmation received was completely at variance with our understanding. He had switched off his hearing aid as he turned to go away, and our "agreement" never had taken place. It was a good lesson for a young purchasing agent not only to be more observant of hearing aids but also to follow through to secure confirmations before misunderstandings become too involved.

Most building contractors operate as owners or proprietors of their businesses. They are in a position to assert themselves. The decisions they make are their own. If they have developed a working relationship with local subcontractors over the years, their decisions will generally be accepted even though they may appear arbitrary to others not familiar with the practices.

The contractor who prepares his own listing of the temporary facilities which he intends to supply for the operation of the job and then sees that all bidders (not only those who are being invited to price the temporary work) are advised to base their proposals accordingly has made a start in the right direction. As the organization grows a bit,

---

* Superior numbers refer to Notes at end of chapter.

others need to follow the same methods to arrive at consistent and comparable conclusions when analyzing subcontract bids.

There are shortcuts in subcontracting work when it becomes necessary to gamble a bit on the details with the expectation that a fair settlement can be arrived at in the future. In taking these steps, it should be recognized that there are recommended procedures which can only be skipped at the risk of becoming involved in the disputes we are trying to avoid.

The contractor whose proposal and general contract is based on a subcontractor's bid that was unusually low—which was received and accepted without thorough investigation—has already started on the road to a controversy. When this occurs, the first step is usually to attempt to secure compliance from a subcontractor who may be equally stubborn. The result could be time-consuming feuding that may continue for years.

Even more unfortunately, a subcontractor's delay in purchasing materials or performing field work can ultimately seriously affect the completion of the building. Before matters reach that stage, it would be well to review the entire situation and see where the error occurred—if in fact there was one.

If the fault was the contractor's, prompt recognition and correction should soon lay the matter to rest. Even if the error was entirely due to the subcontractor, and the contractor had sufficient grounds to accept the bid as offered, a recovery against the subcontractor might not be obtained until long after the job was completed at a higher cost.

The contractor's position here would be far sounder if he had given the subcontractor an opportunity to recheck his proposal and accepted a revision, if necessary. Sometimes there is a failure to spell out sufficiently the details of an understanding that one of the parties considers a binding agreement.

There are contractors and subcontractors in the building business whose analysis of the job specifications is consciously directed toward turning up shortcomings or deficiencies in the specifications. After an award there is pressure for extra work orders. They are gambling; but this type of organization is not general and, while it has its adherents, especially on public work, there is probably less lasting satisfaction and surety of success in this manner of operating.

Another practice reported by one Philadelphia attorney concerned a contractor's dealings with marginal subcontractors: to advance their payments, to guarantee their material bills, and to work out problems with them as they arose. As some of these subcontractors were able to fulfill their contractual obligations, the procedure made it possible to submit lower bids, so that the contractor was "better off" and could secure more work than if he dealt only with more responsible subcontractors. His financially successful career included a 12-hour day, 6 days a week, until he died at the age of forty.[2]

ANOTHER TYPE OF PROBLEM AFTER THE PROPOSAL WAS SUBMITTED

One suit involved an invitation from a government agency to submit a proposal "for the purpose of negotiating a construction contract." In it, the right was reserved "to reject any and all proposals and to negotiate with any proposer." The contractor's proposal had included a statement agreeing to execute a contract "in accordance with the proposal as accepted."

The court held that the negotiation was concluded when the government agency accepted the first proposal. The brevity or length of the discussion did not remove it from the definition of negotiation. Consequently the contractor, who apparently had intended to submit a proposal for further discussion, found himself faced with a suit for damages in the amount of the excess cost of another builder over his original proposed contract consideration of $6,574,825.[3]

The situations that contractors have to face in field operations are many and diverse. Fortunately, trade magazines, construction bulletins, and engineering seminars help acquaint the contractor with solutions that have worked out satisfactorily. Unfortunately, legal problems cannot be avoided as easily.

The arguments in court cases which appear "obvious" to some defendant contractors or subcontractors must appear equally "obvious" to the plaintiff contractor or subcontractor who initiates the suit. All of which points to the treacherous ground on which contractors tread and the need for care in working toward clear understandings without ambiguities that could lead to litigation.

While it does not always follow that a contractor operating most profitably will not adopt an arbitrary attitude and insist on a legalistic interpretation of an agreement, in the long run controversies result from situations where one of the parties stands to lose substantial sums, either through errors in bidding or improper performance by one of the other parties.

Lawsuits are always time-consuming and detract from the productive efforts of a purchasing man whose time might better be spent in careful analysis of costs of other subcontracts still to be awarded. Common sense and calm study will often result in a decision to forget the minor squabbles and chalk up the loss to experience.

The late Joseph G. Fink, an attorney with the firm of French, Fink, Markle and McCallion, who for many years served as counsel to the Building Industry Employers of New York State, edited a series of articles entitled "I Heard of a Case" which appeared in the association's publication, *The Building Industry*.[4]

The articles were also continued for a while by Robert J. Fink and Fred P. Ellison of the same firm. Many of the cases summarized by the author in this chapter were reported in more-extended form in *The Building Industry* articles. Some have been drawn from a continuing series of legal briefs appearing periodically in *Engineering News-Record* and edited by Michael S. Simon. The cases summarized here are only intended to indicate the variety of problems that have reached the courts.[5]

CASES

Some of the cases encountered and the decisions rendered will illustrate the wide range of situations which might exist.

Error in Bidding

PLAINTIFF: Subcontractor
DEFENDANT: Contractor
STATE: Texas
AT ISSUE: Plaintiff had erred in transcribing labor-cost item of $53,951 as $5,395 in establishing contract price of $240,000. Plaintiff sought relief after 6 months had elapsed, and work was substantially complete when error was found.[6]

DECISION: Court in denying relief stated:

> It is not uncommon in the construction field for a subcontractor to submit a bid that is below the cost estimated by the general contractor for the same work and for the general contractor to accept the bid if the subcontractor is reliable and can perform the work. The defendant also took into account that the plaintiff was a local subcontractor and would be able to obtain better prices from suppliers, better cooperation from the unions, more reliable estimates of local costs and incur less in moving equipment, then a contractor based in a distant state.
>
> The mistake was entirely the plaintiff's own, and was not induced in any way by the defendant . . . neither the defendant nor any of its agents knew that the plaintiff was acting under a mistake when it submitted any of its bids. The defendant and its agents had no reason to know that the plaintiff was acting under a mistake either because of the amount of the plaintiff's bid or because of other information they had received from any source.

Another Error in Bidding

PLAINTIFF: General contractor
DEFENDANT: Subcontractor
STATE: California
AT ISSUE: Defendant's telephone bid of $7,131 for paving was submitted to and used by the plaintiff who was awarded the general contract. On the following morning, the plaintiff's engineer was advised of error by the defendant, who refused to perform for less than $15,000. Plaintiff tried for better bids but eventually awarded job to another for $10,948.[7]
DECISION: Trial court, with supreme court confirming, affirmed judgment for plaintiff in the amount of $3,817.

Breach of Contract

PLAINTIFF: Contractor
DEFENDANT: Manufacturer and subcontractor
STATE: New York
AT ISSUE: Contract dated July 25, 1960, which defendant's New York representative had apparently accepted, subject to two minor changes in terms which were interlineated and acceptable to the New York representative. On October 20, 1960, and after the defendant had written that "materials had been entered into production based on plans and specifications obtained from you," the defendant notified the plaintiff that there was no contract. Plaintiff meantime had notified the New York representative that because of the time element and need for the materials, it would not insist on the corrections which had been interlineated.[8]
DECISION: The court held that there was a contract which the defendant had breached by clothing its representative with authority to act. Judgment was awarded plaintiff in the amount of $5,000 with interest from October 20, 1960.

Another Breach

PLAINTIFF: Subcontractor
DEFENDANT: General contractor
STATE: Pennsylvania
AT ISSUE: Plaintiff submitted a proposal which defendant confirmed with its own interpretation, concluding, "Please advise us if this is your understanding." Plaintiff replied, "We accept all terms stated in your letter of July 9, 1955 covering Bessemer Limestone & Cement Co., Job, Bessemer, Pa. Thank you for your business." The letter was sent by registered mail and received by defendant, after which the defendant awarded the work to another concern.[9]

DECISION: The court held that a contract resulted and that the plaintiff had been damaged in the amount of $5,310, and gave judgment accordingly.

A Case of Nonperformance

PLAINTIFF: Subcontractor
DEFENDANT: Contractor
STATE: New York
AT ISSUE: Subcontract provided for certain painting on six houses at a price of $530 each. Prior to completing the work on the first house, plaintiff advised that the coats specified "may not be sufficient to produce a workmanlike job," and that he would not be responsible for the results. Defendant advised plaintiff to proceed under the terms of the subcontract. Plaintiff abandoned job, and the defendant contracted with another to complete. Plaintiff brought suit for $445 for partial completion. Defendant counterclaimed for $500 representing the difference between the price agreed upon the sum paid for completion. Justice court gave judgment for plaintiff.[10]
DECISION: County court reversed judgment, holding:

> There is no justification or excuse for the plaintiff's non-performance or breach of contract. The courts will not consider the hardship or expense or the loss to one party or the meagreness or uselessness of the result to the other. They will neither make nor modify contracts nor dispense with their performance.

Subcontract Interpretations

PLAINTIFF: Contractor
DEFENDANT: Subcontractor
STATE: Massachusetts
AT ISSUE: Deviations from specified masonry construction including omission of parging on inner face of brick wall which was to be tied to concrete-block backup wall; variations in anchoring and cleaning procedures.[11]
DECISION: Supreme court held that tacit approval had been given by contractor in connection with anchoring and cleaning procedures and that these departures from specifications were considered minor; that sequence of work which was approved by contractor contemplated building block wall first before work could begin on brick facing, thereby waiving parging requirements; that the defendant had substantially completed the contract and was entitled to recover *quantum meruit* (reasonable value) even though it had failed to perform in strict accordance with the subcontract.

Written Contract Terms Prevail

PLAINTIFF: Contractor

DEFENDANT: Owner

STATE: Mississippi

AT ISSUE: Claims for extra work performed without written direction from the owner. The dispute involved the demolition and removal of a building and a contract for the construction of a new bank on the same site. The foundations were not to be removed.

The structural engineer, an employee of the owner's architect, told the contractor to remove the old foundations. Contractor notified the owner that this was considered extra work. Lower court allowed contractor to recover the reasonable value of the work.[12]

DECISION: The supreme court reversed this decision, holding that the contract terms which required the owner's written approval for changes took precedence. Court held that the contractor's argument, that the structural engineer was the owner's representative and that trade custom required compliance with such requests, was not binding. "A written contract that is clear will prevail over custom."

Contract Interpretations

PLAINTIFF: Subcontractor

DEFENDANT: Building developer

STATE: Pennsylvania

AT ISSUE: Defendant withheld final payment of approximately $50,000 on a $1,420,000 masonry subcontract, contending that the work had not been performed satisfactorily. He entered counterclaim of $87,000 for overtime wages not incurred, based on contract language stating "This contract is based on working six (6) days a week" and under "Time of Completion" that "The completion to the satisfaction of the architect and contractor shall be had on or before July 1, 1949."[13]

DECISION: The court held that the plaintiff was required to complete the work satisfactorily before July 1, 1949, and that it had to work 6 days a week unless it could finish satisfactorily before July 1, 1949, without working 6 days a week. Jury awarded the plaintiff a verdict of $56,500. The court dismissed the counterclaim.

More Contract Interpretations

PLAINTIFF: Contractor

DEFENDANT: Turnpike Authority

STATE: Kansas

AT ISSUE: Responsibility of highway contractor to redo grading work which had failed after completion, due to unusually heavy rains, under a contract to maintain the grade and drainage until final acceptance of the work.[14]

DECISION: The U.S. Tenth Circuit Court of Appeals affirmed the judgment of the trial court as follows:

> When the contract is viewed by the whole of its parts, it is certainly susceptible of the construction that the parties did intend to specify the circumstances and conditions under which the contractor would be liable to repair and recompact the embankments. Having thus specified the conditions upon which the contractor shall be required to recompact, it is permissible to say, as did the Trial Court, that the plans and specifications, considered as a whole, contemplated that the contractor would be paid for recompacting the embankment which he had constructed according to specifications and which had failed through no fault of his. This construction of the contract is consonant with the evidence in the record to the effect that the parties on the job construed the contract to provide that if the fill or embankment was properly constructed and inspected by the engineers, and if left free draining and soft spots developed, the contractors were to be paid for recomputing the fills and embankments. And it is consistent with the record evidence that this construction of the contract comported with custom and usages in the industry.

Contract Cancellation

PLAINTIFF: Owner
DEFENDANT: Contractor
STATE: Connecticut
AT ISSUE: The contract was executed on the basis of an error which had established the contract price. Contractor, at urging of the owner, prepared an estimate in the amount of $1,450 and executed a contract, discovering that evening that the estimate should have been for $2,210. Contractor offered to perform the work for the revised sum or for as low an amount as any other responsible contractor would charge. The owner insisted on performance at the $1,450 price. Upon contractor's refusal to proceed, the owner awarded work to others for $2,375 and sued the contractor for breach of contract.[15]

DECISION: The court rejected the owner's claim, deciding that although one party's error is not grounds for reformation of the contract, it could offer a basis for cancellation. The court held further that the mistake was so essential and fundamental that the minds of the parties never met; that since the agreement was unperformed and the owner had not been prejudiced in any way, he would not be permitted to gain an unfair advantage over the contractor, even though the mistake was unilateral, and the contractor was entitled to cancellation of the contract; and that although the mistake involved some degree of negligence, it did not amount to a violation of a positive legal duty.

No Contract

PLAINTIFF: Equipment supplier

DEFENDANT: General contractor

STATE: Washington

AT ISSUE: The case of the unaccepted bid and the award for specification items, some of which the plaintiff had quoted on. The defendant had stated: "It looks like you people have the job. However, the purchase order will come out of the field office. We are disappointed that you didn't bid on the hoists." Plaintiff sued for expenses in preparing bid. After trial by jury, the plaintiff was awarded damages, after alleging that by naming the supplier of equipment in the proposal, a commitment had been made. The defendant appealed.[16]

DECISION: Appellate court found that the parties did not arrive at a meeting of the minds and that there was no contract. Court held that a promised purchase order and promised letter of intent did not constitute an acceptance of a bid. These documents were never issued. The court stated that a promise to place an order in the future is "nothing more than negotiation. It indicates an intention to contract at some future date and would not be enforceable."

Offer and Acceptance

PLAINTIFF: Contractor

DEFENDANT: Painting bidder

STATE: California

AT ISSUE: Defendant refused to sign contract based on its revised bid which had been corrected after contractor had notified bidder of a major discrepancy. Contractor used new bid for Navy prime contract, telling the bidder that it had been relied on and that he would receive a subcontract if the prime contract were won.[17]

DECISION: The lower court held that the contractor had relied on a promise that the bidder could not repudiate (promissory estoppel[18]). It held that the contractor had the right to rely on the bid, was unaware of any mistake in the last bid, had accepted the bid within a reasonable time, had not made any counteroffers, and was damaged by the bidder's refusal to perform. The appellate court affirmed the decision, stating that the contractor could recover the difference between the bid in question and the next-lowest bid for performing the work.

Another Offer and Acceptance

PLAINTIFF: Subcontractor
DEFENDANT: General contractor
STATE: Arkansas
AT ISSUE: Plaintiff's telegraphed bid for $3,723 was delivered reading $2,723. Defendant used this amount in preparing proposal and was awarded contract. Notice of error was given defendant by letter.[19]
DECISION: Court held that notice of mistake after plaintiff had materially changed position was too late for plaintiff to correct error that he bore responsibility for.

Loss of Profits

PLAINTIFF: Highway contractor
DEFENDANT: State Highway Commission
STATE: Montana
AT ISSUE: Unearned profits on future work due to a breach of contract in not providing right of way. Because of being tied up on job for over 2 years, the contractor lost bonding capacity and thereby lost future profits.[20]
DECISION: Supreme court held that general principles of law precluded recovery on unearned profits. It also noted that "future unearned profits might be allowed when the contractor can prove with reasonable certainty what foreseeable profits would have been earned but for the breach." Jury awarded $78,000 which the court stated was reasonable.

More Lost Profits

PLAINTIFF: Assignees of roofing subcontractor
DEFENDANT: Developer
STATE: California
AT ISSUE: Scheduling of roofing work on 936 houses with contract price of $355,000. Defendant represented that work would begin within 3 weeks (by April 1). Defendant several times advised that work would start in 2 or 3 weeks. On May 24, plaintiff commenced action on the theory that the defendant's continued delay constituted a breach of contract.[21]
DECISION: As no time had been specified for commencement of work, the court found that a reasonable time had expired on May 17; there had been a breach of contract rendering the defendant liable for loss of profits the roofing contractor would have made on the job.

Recovering Damages

PLAINTIFF: Subcontractor
DEFENDANT: State of New York
STATE: New York
AT ISSUE: Claim for damages incurred due to a 2-year delay in completion of general construction.[22]
DECISION: Court concluded that the State had failed to make "serious and substantial efforts to progress and coordinate the work"; the subcontractor was permitted to recover his damages. State failed to enforce general contractor's obligation to supply hoists and elevators.

Damages—Nonrecoverable

PLAINTIFF: General contractor
DEFENDANT: State of New York
STATE: New York
AT ISSUE: Claims for damages sustained due to delays caused by plumbing subcontractor for State, not attributable to weather or unreasonable duration of work.[23]
DECISION: Court held that the State had made sustained efforts to compel plumbing subcontractor to complete work on time and had refrained from canceling contract or requiring surety to complete because it concluded that neither of these actions would achieve the desired goal—prompt completion of the work. The State was held not liable for damages to the plaintiff.

Failure to Coordinate the Work

PLAINTIFF: Subcontractor
DEFENDANT: State of New York
STATE: New York
AT ISSUE: Failure of State to control program of work and coordinate various contractors on mental health building with consequent delay to plaintiff (plumbing subcontractor) whose contract called for completion by January 1, 1955.[24]
DECISION: Court found that the State failed to see that the work progressed properly and was properly coordinated, and had unreasonably failed to supervise, coordinate, and see to the progress of the work.

Damages allowed: Additional fire insurance due to lapse of time attributable to delay resulting from breach of contract; additional costs due to increase in wage rates beginning May 1, 1955.

Damages disallowed: Supervisors' trips and expenses after January 1, 1955; additional foreman's wages.

Another Failure to Coordinate the Work

PLAINTIFF: Subcontractor
DEFENDANT: State of New York
STATE: New York
AT ISSUE: Failure of State to properly supervise progress of defaulting general contractor who undertook to complete work on same completion date as plaintiff.[25]
DECISION: Appellate court held State liable for damages occasioned plaintiff.

Recovering Damages

PLAINTIFF: Subcontractor
DEFENDANT: Insurance company
STATE: New York
AT ISSUE: Responsibility of insurance company to reimburse subcontractor for damages of $4,913 it paid to owner and tenants of apartment building in which rain damage occurred during a roof resurfacing job. Insurance company disclaimed responsibility on the grounds of faulty workmanship and on grounds that damaged property was in the care, custody, and control of the subcontractor.[26]
DECISION: Appellate division of supreme court held that the policy covered loss; that occurrence fell within definition of "accident"; and that the property damaged was not in the care, custody, and control of the subcontractor.

Arguments over Progress Payments

PLAINTIFF: Contractor
DEFENDANT: Subcontractor
STATE: Maine
AT ISSUE: Subcontract involved delivery and installation of reinforcing steel for six buildings. Subcontractor alleged failure to make progress payments and to coordinate progress. Subcontractor left job just prior to finishing its work on a date after general-contract-set completion date.[27]
DECISION: Court found that the contractor had paid for steel "delivered and erected." Subcontractor had billed incorrectly on the basis of delivery only. Court ruled that the subcontract did not make time of the essence and that the subcontractor had breached the agreement and was liable for the contractor's damage.

Two Points in Contention

PLAINTIFF: Subcontractor
DEFENDANT: Contractor
STATE: Washington
AT ISSUE: Effectiveness of contract for $1,051,453 even though agreement was not reached on one item, later established as costing $9,355.[28]

DECISION: Court assumed that the parties, if they had agreed, would have compromised at a figure between zero and actual cost, and awarded the subcontractor $4,677.

AT ISSUE: Contractor further refused to pay 10 percent withholding, contending that the subcontractor had been responsible for delays in completing the general contract. Subcontractor had agreed to a completion schedule which the contractor contended was to permit completion of general construction.

DECISION: Court held that there was nothing to obligate the subcontractor to perform work at its own cost and expense in the event the agreed-upon schedule did not permit the contractor to meet his time requirements.

More Double-Trouble

PLAINTIFF: Contractor
DEFENDANT: Building owner
STATE: New York
AT ISSUE: Defendant advised plaintiff that he was to arrange for the demolition of an existing building, take bids, perform preliminary quantity survey work, let subcontracts, and supervise construction for a 20-story apartment building, for which a $45,000 fee would be paid. Following receipt and review of bids and extended conferences by the plaintiff, the defendant advised that it had been decided not to construct the building and that the property would be sold. Afterward, the defendant owner hired another contractor to complete the building. The owner testified that he had never engaged the plaintiff and that the plaintiff's services had been volunteered in the expectation of being engaged as the contractor. Owner claimed that by the New York Statute of Frauds an oral agreement was void, if, by its terms, it would not be performed within 1 year from the time it was made. Secondly, the owner claimed that the architectural and engineering provisions of the state education law barred enforcement of contracts for architectural and engineering service except by licensed architects and engineers.[29]

DECISION: The court found that the work was clearly capable of being performed within 1 year. On the second matter, the court overruled this defense on the grounds that the work performed was not prohibited by this provision and that the work was subject to the approval of a registered architect or engineer in the employ of the owner. The court found that the contractor had completed 40 percent of the work, for which he was awarded 40 percent of the fee. One expert testified that the preliminary work constituted 60 percent of the price with 40 percent for supervision. The owner's architect estimated preliminary work to be valued at 40 percent with 60 percent allowed for supervision.

ARBITRATION

A helpful reference on this topic entitled "Delays and Disputes in Building Construction" outlines a variety of problems affecting general contractors and their relationship with subcontractors, owners, architects, and sureties. The arbitration process is described in detail, as are the applications and benefits of resolving disputes expeditiously.[30]

"Arbitration is contractual in nature"; i.e., "it is possible only if the parties can be deemed to have agreed among themselves to its use."[31]

Where there is an agreement, on the recommendation of an attorney familiar with local construction practices, to include an arbitration clause in the subcontract form, the basic wording might well state the following:

"Any controversy or claim arising out of or relating to this subcontract, or the breach thereof, shall be settled in accordance with the Construction Industry Arbitration Rules of the American Arbitration Association, and judgment upon the award may be entered in any court having jurisdiction thereof."

The phrasing of matters being submitted to arbitration should be precise. It is possible to include several actions combined and consolidated in one hearing. Generally speaking, the arbitration awards are final and binding and are subject to the enforcement of the courts.

As an example, in a case involving a Connecticut municipality which questioned an award as not being definite and final, the supreme court ruled that an award that answered the questions submitted cannot be challenged because it does not contain the arbitrator's reasoning.[32]

Arbitration procedures are becoming more standardized; the number of cases in the construction industry is increasing as the techniques become more familiar, as the cases can be scheduled faster, and, for convenience' sake, can be held at the construction site. Available arbitrators who are knowledgeable about the construction industry are usually assigned.

At the same time and despite the popularity and great number of procedures due to the economy, speed, and expertise in the arbitration, it must be recognized that some of the decisions lack harmony, especially with respect to the sureties.[33]

In 1969, a study was made of 30 typical construction disputes in which arbitration awards were rendered. The most recent cases from the American Arbitration Association's New York region were assessed.

> Experience and prior studies indicate that this sample is typical of construction cases in New York and elsewhere through the AAA system.
>
> The average size of the claimant's demand was $5,775. The median was considerably below that figure, $4,770. The smallest was for $315 and was denied. The largest claim was for $20,590 and was granted in full with interest.[34]

Even though some court and arbitration claims involve very large sums, for the most part only comparatively modest amounts enter into claims. Certainly, more care in negotiating and a greater element of fairness on the part of both contractor and subcontractor could go far in reducing the number of controversies which ultimately reach the courts or arbitration panels.

Our own experience leans toward the omission of the arbitration clause in the subcontract form which the general contractor prepares, to avoid the tendency on some subcontractors' parts to request arbitration on minor matters which normally can be resolved in the day-to-day operation of the job.

While recognizing the practicality of the above, should both parties determine that an arbitration clause is desired, it should be drafted and added by an attorney familiar with local practices and should clearly specify the types of potential problems to be arbitrated.

NOTES—Chapter 8

1. Lyle E. Treadway and Gordon Burt Affleck, "Legal Aspects," in *Guide to Purchasing*, National Association of Purchasing Agents, Inc., 1967.
2. Edward H. Cushman, "Contractor Obligations under the Miller Act," *Constructor*, May 1965.
3. *U.S. v. McShain*, 258 Fed.2d 422.
4. During the years from 1948 through 1965, over 170 cases which involved contractors' relations with subcontractors, owners, and others were reported in *The Building Industry* articles.
5. G. A. Leonards (ed.), *Foundation Engineering*, McGraw-Hill, New York, 1962, includes a section on "Legal Aspects of Earthwork and Foundation Engineering" by I. Vernon Werbin, in which reference is made to 50 cases involving this general subject.
6. *Poley-Abrams Corp. v. Chaney and James Construction Co.*, 220 F. Supp. 401.
7. *Drennan v. Star Paving Co.*, 332 P.2d 757.
8. *New York Law Journal*, vol. 149, no. 122, June 25, 1963.
9. *Youngstown Steel Erecting Co. v. MacDonald Engineering Co.*, 154 Fed. Supp. 377.

10. *A. Berger Sons, Inc. v. Mogul Construction Co., Inc., New York Law Journal.* Apr. 2, 1954.

11. *Hayeck Building & Realty Co., Inc. v. Turcotte,* 282 N.E.2d 907 (1972).

12. *Citizens National Bank of Meridan v. L. L. Glascock, Inc.,* 243 So.2d 67 (1971).

13. *John B. Kelly, Inc. v. Aronimink Village Apartments, Inc.,* U.S. District Ct. Pa., July 26, 1954, 122 Fed. Supp. 905.

14. *Kansas Turnpike Authority v. Abramson,* 275 Fed.2d 711.

15. *Geremia v. Boyarsky,* 107 Conn. 387, 140 A. 749 (Conn. 1928).

16. *Merritt-Chapman and Scott Corp. v. Gunderson Bros. Engineering Corp.,* 305 D.2d 659.

17. *H. W. Stanfield Construction Corp. v. Robert McMullen & Son., Inc.,* App. 92 Cal. Ptr. 669 (1971).

18. Promissory estoppel makes a promise binding upon the promissor *if* he could reasonably expect the person to whom the promise was made to act (or not to act) in a certain way, *and* the person did indeed do so, to his disadvantage because of the failure of the promissor to live up to his promise.

19. *Buchanan v. Thomas,* 320 S.W.2d 650, in *New York University Law Review,* "Annual Survey of American Law," February, 1960.

20. *Laas v. Montana State Highway Commission,* 483 P.2d 699 (1971).

21. *Stark v. Shaw,* 317 P.2d 182, Cert. denied 356 U.S. 937.

22. *Forest Electrical Corp. v. State of New York,* 198 N.Y. Supp.2d 600.

23. *Norelli & Oliver Const. Co. v. State of New York,* Third District, Oct. 17, 1968.

24. *Snyder Plumbing & Heating Corp. v. State of New York,* 198 N.Y. Supp.2d 600.

25. *Weil Plumbing Corp. v. State of New York,* 267 App. Div. 247.

26. *Rex Roofing Company v. Lumber Mutual Casualty Insurance Co.,* 280 App. Div. 665.

27. *Drew Brown Limited v. Joseph Rugo, Inc.,* 436 F.2d 632 (First Cir. 1971).

28. *Purvis v. U.S.A. for use of Associated Sand & Gravel Co., Inc.,* 344 F.2d 867.

29. *Crystal v. Denberg,* N.Y. Sup. Court, 237 N.Y.S.2d.

30. National Construction Industry Arbitration Committee, "Delays and Disputes in Building Construction," New York, 1970.

31. Robert Coulson, "What is Happening in Construction Arbitration?" American Arbitration Association, 1969.

32. Patrick E. Hartigan, "Effect on Surety of Arbitration Award Adverse to Principal and Surety's Right to Participate, and the Consequences thereof, in the Arbitration Process," *The Forum,* July 1969.

33. Gerald Aksen, "Resolving Construction Contract Disputes through Arbitration," *The Arbitration Journal,* vol. 23, no. 3, 1968.

34. Patrick E. Hartigan, op. cit.

PART FOUR

Department Operation

CHAPTER NINE

Policies, Procedures, and Training

The construction company—whether it involves a one-man organization where the boss secures the contract, does the buying, and supervises the work or a progressively larger firm with several departments—is primarily devoted to giving service. The purchasing or subcontracting function should be one of the links that bind the company together into a profitable operation. It can chart the course for a smoothly running company. A lack of business training in this phase can be disastrous, even before the first contract is secured.

The company's policies are a reflection of the character of the individual leading it; its reputation is seldom acquired overnight, but rather is a result of day-in-and-day-out fairness coupled with a sound sense of business values. There needs to be an ability to communicate effectively with others, which comes from entering into the mainstream of the business and civic life of the community.

Every company, in whatever business, needs to have its established policies and procedures for operating. At the same time, one does not start by writing a manual, and then go into business. Rather, it becomes a case of summarizing the decisions that have been made over a period of years and then of grouping them in a form to enable newcomers in the organization to learn.

Most companies, and certainly all companies in the construction field, are in business to show a profit. For the construction company to continue to operate, the bids submitted must be high enough to include this allowance but low enough to get the job.

The "strategies," if you will, to secure work should be based on selectiveness in deciding which projects appear attractive, in analyzing the competition, and in determining how best to utilize the firm's financial resources, manpower, and reputation to accomplish its objectives. One of Webster's definitions of strategy as the "art of trickery," or getting the better of an adversary, has no place in the fundamental policies by which reputable contractors operate.

COMPANY ORGANIZATION

There are many variations followed in setting up the operations for construction companies. Commonly, the sales or contract department will be the basic arm of the administration, supported by estimating and purchasing. The operations department frequently handles the field construction and also coordinates the purchasing and engineering functions after a general contract has been secured. Accounting, costs, and personnel serve as supporting departments for most companies.

Many companies place the responsibility for securing the contract and for its administration in a project manager or project executive who coordinates the sales and estimating efforts as well as the later contract performance.

STARTING UP

In the examples given, it has been assumed that the company has been operating and that there have been some jobs estimated successfully and completed—in other words, there is a going business.

On the other hand, if it's a case of weighing the possibilities of getting started, a knowledge of some of the office fundamentals is highly desirable. Hiring personnel, buying office equipment, and handling accounting and correspondence are only a start in the process. To operate efficiently, it will be necessary to have a basic knowledge of the hundreds of small decisions that need to be made in getting organized. If ready answers are available, based on experience, operations will go more smoothly.

Presumably the new contractor knows the ropes from the field and estimating angle. He also needs to know where to turn for a good grasp of office fundamentals and to have thought through the comparative situations in his own case. While businesses have been started with the "profits" consisting of a journeyman's wages, or its equivalent, and the contractor's wife handling the office work, a slower start with one or two associates handling other areas of responsibility and competency is more apt to result in a firmly established and profitable operation.[1,*]

SPECIFICATIONS—THEN AND NOW

Over the years there have been many efforts to standardize specifications for all areas of construction. Architects' and contractors' checklists have been revised and annotated by the specification writers, and the likelihood of major omissions has been lessened on the standard elements in a building.

At the same time, there have been so many new techniques and products in the architectural and construction field that call for new items on the checklist that one must be alert to arrive at a construction agreement where all the contractor and subcontractor relationships have been considered in advance. Even then, with some of the nationally known manufacturers of building products, there are occasional offerings that should arouse skepticism on the part of the contractor and buyer, to avoid being responsible for the corrections likely to be required. There may be grounds for not bidding projects where the risks outweigh the rewards.

Some of the larger construction companies assign one individual to investigate and report on technical aspects of their projects: the causes of roofing failure and glass breakage; the problems with soil drainage systems; or the deficiencies in manufacturing cement, plaster, or paint that resulted in failures. As this information is accumulated, it forms the basic core of knowledge that can assist producers in their particular industry.

In Chapter 3 we noted that situations have developed in which the services of an appraiser were needed and that some contractors with those capabilities have been able to combine their estimating and engineering services to considerably augment their annual volume.

Building-construction projects are the testing grounds for hundreds of new products that cry for research and development reports. Some of the larger contracting concerns are prepared to undertake investigative studies for building-material producers and can offer counsel as a result of field experience. It is a new field requiring technically trained personnel operating as a completely separate section of a contractor's organization to prevent any conflict of interest.

* Superior numbers refer to Notes at end of chapter.

COMPARISON WITH
INDUSTRIAL SUBCONTRACTING

The procurement function, whether it applies to buying materials or awarding subcontracts, is performed to some degree in all companies. It is interesting to compare the decisions which the construction contractor will have to make and the basic similarities with the "make or buy" factors encountered in the industrial management field,[2] as noted below:

Time Available The experienced subcontractor will be able to bring his skills into action quickly and start work promptly. Presumably he also has the necessary equipment available for the work.

Location The subcontractor operating in the area can be effective as long as transportation and liaison are efficient and do not create major obstacles.

Comparative Costs Initially, without a good background of cost information, the contractor cannot compare his costs against those for trades such as formwork, concrete, masonry, or carpentry which he might perform with his own forces. As he accumulates more data and observes subcontractors' procedures, he will be able to make valid comparisons.

Relative Prices So long as the subcontractor's estimates are reasonably close to the contractor's estimates, the tendency will be to use subcontractors and lessen the risks which might be encountered. The contractor can devote his energies to other areas where his own experience may pay off better.

Product Consideration If the contractor's organization is fully occupied and if the subcontractor gives evidence of ability to perform, the subcontracting method will undoubtedly be favored since it gives greater flexibility. It tends to lose its advantages when the contractor must provide for the same kinds of work on each project, e.g., miscellaneous carpentry items involving protection, fencing, and general cleaning tasks around the job.

Market Factors As long as there is a reasonable number of subcontractors, there is minimum risk in placing confidence in the subcontracting method. Basically it avoids the immediate need of working capital for payrolls and the purchase of new equipment by utilizing the subcontractor's. It avoids an overexpansion in the equipment field. This expansion may be of dubious value later, and might prevent taking advantage of subcontractors' attractive bids in the future when they are not fully engaged. The contractor should avoid committing all his work to one subcontractor, where there is any choice.

All the factors mentioned raise questions that can be answered only in the light of the particular situations involved. However, the comparisons are worth considering. One major difference is the tendency of most large industrial concerns to have buyers for one or more groups of items. Construction firms usually have one individual or specialist contract for all the 30 or more trades on a project, whether it be the structural, architectural, mechanical, or electrical system.

In our analysis, especially with smaller buildings involved, the examples include all trades. It becomes a matter not of knowing all there is to know about excavation and the structural, finishing, or mechanical trades, but rather of being able to recognize the accuracy and complete- ness of the bidding information and arrive at a complete subcontract package "with a string around it." The contract should meet the needs of the field organization at a price, fairly negotiated, which will permit both contractor and subcontractor to remain in business and should lay the groundwork for future efforts.

RELATIONSHIPS WITH SUBCONTRACTORS

Most purchasing people, while serving as employees of a contractor, find that they need to establish close business relationships with the subcontractors and salesmen, to maintain their friendship and knowledge in order to be able to rely on them for economy, quality, service, and progress. At the same time, it is important that those purchasing do not place themselves in a position which permits suspicion of their ethics and motives. Any substantial stockholding by purchasing agents in a subcontractor's or vendor's company that would indicate a possible con- flict of interest should be avoided.

By virtue of his title, the purchasing agent serves as an agent of the company he represents. He needs to have a sufficient understanding of the relationship between the company and its subcontractors and of the consequences of the acts performed in the company's name.

The acts of those buying, as agents, are binding on the company within the limits of the authority given. In order to avoid personal liability, the buyer should always make clear to the seller that he is acting as agent.

There is personal liability involved if the buyer:

1. Makes a false statement concerning his authority, with intent to deceive, or if his misrepresentation has the natural and probable conse- quence of being misleading

2. Performs without authority a damaging act, even though believing he has such authority

3. Performs an illegal act, even though with the authority of his employer

Once a subcontract is made, both parties are legally obligated to perform in accordance with its terms, unless, of course, both agree to a change.

RELATIONS WITH OTHER DEPARTMENTS

As the operation grows from a one-man function responsible for the subcontract agreements, change orders, and approvals to one where another handles the change estimates and the liaison with the architect and owner, the need for circulating information increases. It is here that much thought is necessary to avoid building up a paper empire which, while helpful, may also be unnecessary if the data is reasonably accessible to others.

Usually one individual (frequently the project engineer on the larger jobs) in the contractor's organization will be assigned to handle the cost estimates for plan and field changes and their approval by the architect and owner. Commonly all data in connection with a particular revision are given a change estimate (CE) number. Requests for approval can be forwarded as "A" letters. After approval, a change order, either addition A-# (Fig. 16) or deduction D-# (Fig. 17), is prepared

---

A. B. C. CONSTRUCTION COMPANY

CHANGE ORDER NO. A-2   **SUBCONTRACT CHANGE ORDER**

ADDITION

BUILDING A. B. Jones Building, Fremont, Ohio   NO. 57

TO CONTRACT PRICE

SUBCONTRACT WORK ___Concrete___

THE TERMS AND CONDITIONS OF THE ORIGINAL SUBCONTRACT FOR THE ABOVE WORK AT THE ABOVE BUILDING SHALL GOVERN THIS CHANGE.

November 15, 1974

T.R.F. Concrete Company
Marion, Ohio

Gentlemen: Re: Approval Letter A-12   CE #22

    This will confirm authorization for you to perform additional work on the above job as listed below:

Your proposal dated Nov. 1, 1974,

    For changes in elevator pit construction

    per architect's revised drawing A-2 dated

Oct.15, 1974. $134.00

    In consideration of the above, your subcontract is increased by the sum of One-Hundred Thirty-Four Dollars ($134).

Very truly yours,

J.J. Smith

Purchasing Agent

FIG. 16   Change Order—Addition.

A. B. C. CONSTRUCTION COMPANY

CHANGE ORDER NO. D-3   SUBCONTRACT CHANGE ORDER

DEDUCTION   BUILDING A.B. Jones Buidling, Fremont, Ohio   NO. 57

FROM CONTRACT PRICE   SUBCONTRACT WORK   Electrical

THE TERMS AND CONDITIONS OF THE ORIGINAL SUBCONTRACT FOR THE
ABOVE WORK AT THE ABOVE BUILDING SHALL GOVERN THIS CHANGE.

January 5, 1975

N.L.D. Electrical Company
Logansport, Indiana

Approval Letter A-19
Change Estimate CE #27

Gentlemen:

This will confirm authorization for you to make changes as
defined by architect's revised drawings dated October 15, 1974, for
which you quoted a credit of $575 in the proposal dated November 5, 1974.

It is understood that the specified excavation work involved will
be performed by others, consequently the credit was increased by $100 as
agreed.

In consideration of the above, the subcontract is reduced by
the sum of Six Hundred Seventy-Five Dollars ($675).

Very truly yours,

J. J. Smith
Purchasing Agent

**FIG. 17   Change Order—Deduction. This form is usually printed in red
to distinguish it from the Change Order—Addition.**

by the purchasing agent with the amount entered on the subcontract
record cards. These changes are recorded in the subcontract ledgers
by the accountant.

Subcontract information letters (Fig. 18), where no change in price
is involved, commonly are prepared for the record to confirm architect's
approval of changes.

Efforts should be made to minimize copying by attaching copies where
necessary, but endeavor to avoid duplication of stenographic efforts
which become extensive when many change orders are involved on a
project. The average purchasing department with one secretary for
each buyer will be fully occupied with the preparation of bidding documents, subcontracts, and change orders and with the daily correspondence of an office. No time will be available for duplication of routine
copying that can be avoided.

Standard accounting procedures recommend that no person should
be in complete control of an important part of a transaction. Office

---

A. B. C. CONSTRUCTION COMPANY

SUBCONTRACT INFORMATION

BUILDING_____A. B. Jones Building, Fremont, Ohio NO.___57___

SUBCONTRACT WORK_____Roofing_____

March 10, 1975

H.A.C. Roofing Company
Akron, Ohio

Gentlemen: Approval letter A-25   CE #52

      The subcontract dated December 30, 1974, is amended

and clarified as follows:

      The architect has approved the use of roofing

materials manufactured by A.B.C. Company in lieu of

specified materials by D.E.F. Company.

      No change in price is involved.

                Very truly yours,

                J. J. Smith
                Purchasing Agent

FIG. 18   Subcontract Information Letter.

procedure can be improved through a system of checks and balances wherein two or more methods might be developed along separate lines or carried on by different persons. The result developed by each procedure affords a check or basis of appraisal against the other. The time-keeping records on payroll work can be applied to material cost analysis. The field superintendent, when the work is not subcontracted, can apply the estimated quantities of concrete for a certain area against the installation costs for the concrete as appearing on the delivery tickets.

The possibility of error is greatly increased if no basis of comparison is readily available. When the work of any one person is subjected to review by a second under a normal routine, there is added assurance of accuracy up to that point. There is also the greater probability that the first individual will function more effectively when he knows that his work is being reviewed. The exception of course would be the contractor himself, but, here again, experience will confirm the advantage if some other individual checks the "arithmetic" whenever computations are involved or if a board of directors confirms the thinking when major decisions are imminent.

One form useful in evaluating purchasing performance lists the subcontractors' bids as originally received together with the final prices which each subcontractor eventually determined as his best bid. It permits the purchasing agent to review his own performance objectively, compared to the estimate, and also indicates to the department head the tightness or softness of the subcontractors' current pricing.

During the past decade we have entered the computer age when data processing has proved useful in several phases of building contracting including not only payroll and timekeeping records, issuance of purchase orders, and complex estimating but also structural design.

It can apply especially to estimating structural and mechanical work where well-defined units are tied in closely to labor requirements. Once the material takeoffs are completed, the estimate becomes a function of the material prices and wage rates. There is need for a broad perspective to take advantage of the new methods.

The possibilities should not becloud the necessity for a personal understanding of how bids are prepared and of how work is subcontracted and laboriously fitted together in buildings throughout the country. Without this basic knowledge, the contractor is like a ship's captain who relies wholly on his radar without posting a lookout to serve when the need arises.

PURCHASING PERSONNEL

Personnel development, whether it be for the purchasing function or some other phase of construction management, has many characteristics and qualifications against which an individual can measure himself. Certainly, one possessing all the highest attributes listed in an evaluation would be a paragon of virtue. Without assigning priorities, each attribute is important in its way. Each takes individual analysis to bring about improvement as well as consultation with another to reach one's maximum potential.

Character: The moral qualities or traits such as integrity, sincerity, and moral courage

Emotional Adjustment: The faculty to adjust to realities and direct one's energies and inner drives into productive channels

Stability: The ability to work effectively under pressure

Ambition: The desire or eagerness to achieve some worthy goal or the goal itself

Drive: The energy used to tackle the job in hand or to achieve one's ambition

Initiative: The faculty for initiating action in the absence of detailed instructions or in the face of obstacles

Motivation: The incentive such as desire for respect, security, or salary which moves an individual to work effectively for the company

Creativeness: The ability to formulate new ideas which is manifested in the development of new and better ways to perform a given task

Resourcefulness: The capacity for improving or finding expedients to overcome obstacles

Judgment: The ability to evaluate a situation or set of facts and arrive at a sound conclusion

Delegation: The act of appraising a subordinate's capabilities; assigning him a task with the instruction, assistance, authority, and freedom of action which he will require; and performing the follow-up necessary to see that the assignment is successfully carried out

Leadership: The ability to motivate others and to gain their acceptance of one's beliefs and objectives

Responsibility: The degree of accountability which an individual assumes for the results of his performance

Communication: The ability to express oneself clearly and the faculty for keeping others properly informed on matters within their sphere

Planning: The visualization of the various aspects of an assignment and the systematic development of a sound plan of action for its accomplishment

Organization: The ability to effectively implement a planned course of action and to integrate the various components

Enthusiasm: The quality that brings with it the challenge of adventure and excitement in one's work

Productivity: The degree of effectiveness with which an individual uses his time and energies in terms of quality and quantity of output

Teamwork: The ability of an individual to work effectively with others for a common goal through the subordination of his personal desires

To a large extent, the same factors that will help him improve performance in purchasing also apply in other phases of construction work, for instance:

The willingness to accept criticism and act upon constructive suggestions from his superiors, subordinates, or associates.

The ability to be his own "spark plug" to motivate himself and others to action and decision.

The ability to deal effectively with a wide range of personalities in his many daily contacts.

The ability to be flexible and adjust to new courses of action when necessary without loss of effectiveness.

The ability to communicate clearly in both writing and speaking.

The ability to maintain a feeling of equality in the daily business contacts with technical men and to command their respect through practical knowledge of the construction business.

Fair dealing, integrity, service, progressiveness, and friendship will all help enhance the purchasing agent's reputation as well as that of the organization.

Last, but by no means least, interest in the work, mechanical aptitude, analytical ability, and reliability all play their parts in the making of a purchasing man who can best serve the contractor and through him the client for whom they work.

CONTINUING EDUCATION

College educators report that 50 percent of the engineering graduates need refresher courses after having been out of college for 10 years. To keep abreast of the times, there is need for continuing education in the form of extension courses or attendance at engineering schools and for review of available trade publications. Observing other contractors' operations may be a surefire way to avoid some troubles as well as to get new ideas.

Study of a general construction handbook covering aspects of building design and construction ranging from the selection of the architect through the stages of construction management, including estimating, insurance work, and other supporting activities, can be helpful and informative.[3]

There are many examples of successful contractors who started their business without going to college. The common denominators, and they are not limited to contracting, are the qualities of enthusiasm, integrity, and tenacity and the development of an effective means of communication to combine with the technical experience which the contractor possesses.

TRAINING IN PURCHASING

The individual's training in a smaller company can be expedited if the new man is assigned to work under the direction of one actively engaged in a particular project. Invitations to bid to prepare and issue, selection of subcontractor bidding, quotations to record and analyze, and questions for which answers must be found and transmitted to the subcontractor bidders are all time-consuming. The new man will find himself well occupied and at the same time increasingly will get the feel of the office. This "big brother" technique, as it is sometimes called, brings with it an insight into office procedures and the practical day-to-day business experiences in the trade.

There are advantages in having those engaged in receiving subcontractors working in the area where the "give and take" of negotiations with many subcontractors registers with the buyer, beyond his particular project. He needs to know which subcontractors are being dealt with on other jobs (and their performance on those jobs) before awarding additional work. Perhaps there are open items on claims or backcharges that can be settled quickly before closing out new subcontracts.

If the contractor's staff is large and involves several departments, it will help the general efficiency if thought is given to rotation at intervals to familiarize the various individuals with the overall operation of the business.

To a considerable degree, the purchasing function in a construction company may involve buying a service, for instance the concreting in a building, and at the same time may involve "selling" the subcontractor bidder on the advantages of taking the subcontract. This ability to wear two hats can be developed through the knowledge which comes from field experience, from frequent dealings with subcontractors, and from a basic appreciation of what represents a fair price for the work in question. It helps to be able to know what to say, how much to say, how to say it, and when to say it.

Fundamentally, the success of most construction companies can be traced to team effort in carrying out the various projects under the overall direction of an able leader. The selection of the right compatible personnel for a job and the coordination of the individuals involved becomes of great importance to the contractor.

The experienced buyer will know when to press an advantage for a better deal as well as when to conclude a negotiation. A too-low price that brings later recriminations on the part of the subcontractor and risks a default is never a good investment. This sixth sense in bargaining not only tends to avoid sharp practices and efforts to beat down price levels but also is based on a knowledge of the reasonable pricing for the work in question that has been tested over the years.

FIELD PURCHASE ORDERS

Purchase orders for supplies and miscellaneous items used in the field normally are prepared by the material clerk for the superintendent's signature and usually are issued as confirmations (not to be duplicated) to verify the processing of payments. Where the jobs are larger, with many requisitions, these are usually numbered and prepared by several individuals. If the amount is small, say less than $25, they can be matched up with the material receipt and numbered invoices as received, and the purchase order dispensed with. Frequently field order forms carry some code to distinguish them from those issued by the home-office purchasing department.

On larger projects with governmental or owner's approval required on purchase, some requisitions, prepared by a storekeeper and signed by the superintendent to conform to contract terms, are often transmitted to the purchasing department for purchase. It is usual to set some kind of limit, perhaps $500, above which no purchase is to be made without the owner's prior approval. This tends to make the storekeeper feel responsible for his inventory. Any efforts to split requisitions or orders to evade the $500 limitation should be noted and corrected.

Modifications of purchase orders usually are issued to correct major quantity revisions but normally are not prepared for minor underfills or overfills.

CHECKLISTS FOR SUBCONTRACTING

In Chapter 4 some of the factors influencing pricing of subcontracts were noted. The list can be extended with items such as those shown below to form the first stage in establishing guidelines of what may eventually become a manual for the department. Apart from the advantages it offers in recalling points to consider in buying, the manual can also standardize procedures and encourage thoughtful analysis of the terms specifications may not have covered which are peculiar to the contractor in question.

There are many alternatives in using a purchasing checklist, any one of which can be acceptable if there is mutual agreement when subcontracting. The point to avoid is having no understanding and being faced with an arbitrary charge or settlement by a subcontractor.

The list can be accumulated in a loose-leaf book to form a personal account for the buyer with each item recalling some specific instance from his experience. Because of the variety and number, there are advantages in separating them into groupings, such as:

A. Operating conditions, temporary work, plant installations, job equipment, and construction scheduling
B. Ways to reduce costs
C. Establishing costs of changes and methods of handling changes
D. Special subcontract terms used in preparing agreements
E. Specification deficiencies and information needed for negotiating

For example, under heading A, does the acoustical subcontractor need shop drawings to install his work properly, and are they specified? Is the job complicated enough to require them?

Is the acoustical subcontractor or electrical subcontractor responsible for the ceiling layout? Have all samples including adhesive been approved by the architect? If a failure occurs even though a top-quality material is used, if its use was not specifically approved, requirement for making replacement will be the subcontractor's and responsibility of the contractor.

Under heading B, would shipments made directly from the mill be more economical than fabrication in the local metropolitan area?

Could one of the field coats of paint be applied on the steel sash before erection?

Under heading C, secure breakdowns of a sample estimate to show the proporation of foreman's time with respect to the mechanics' hours with percentages for fringe benefits, insurance, overhead, and profit.

Secure hourly rates for drafting charges on structural-steel and miscellaneous iron subcontracts which need not necessarily reflect industry standards.

Secure hourly rates for painting extras including materials, supplies, use of dropcloths and ladders, etc.

Secure hourly rates for plaster patching including materials.

Establish basis for cost-plus plaster patching with some area noted above which unit prices would apply.

Are doors and windows included in estimating takeoffs for extra wall areas?

What effect does room size have on the unit prices for flooring, acoustical ceilings, etc.?

What types of work will be performed without comeback charges?

Are percentages allowed for overhead and profit on premium time?

Has bulking factor been considered in computing excavation, e.g., 9 cubic yards in place could represent 10 cubic yards by truck measurement?

Have erection costs been established for different types of electrical fixtures?

Under heading D, has an agreement been reached on overblasting and rock leveling of footings in excavation subcontract?

If a temporary stoppage of the work occurs, what understandings are there affecting equipment rentals, permanent personnel, etc.?

Under heading E, has general excavation been properly described and classified? Would the "lower" boiler room and larger recess in rock faces be classified as "general rock" or as "trench excavation" at a higher unit price?

Is excavated material acceptable as backfill?

What provision is made for backfilling if it is partly due to overexcavation?

While architectural changes involving additional sprinkler heads will warrant an extra, is the subcontractor responsible for installing additional heads necessitated by the Factory Insurance Association's or National Association of Fire Underwriters' examination of plans?

MANUALS

A manual gives a basis for the expectation that the organization's policies and procedures will be followed in the various offices of a company and that subcontractors will be afforded the same treatment wherever they operate.

For the larger building concerns, manuals, whether they are for time-keeping, engineering, operations, or purchasing, are a must, so that trans-ferred staff and new personnel will have similar standards to follow. Revisions and amendments will come as a matter of course with several persons submitting ideas on the updating. Having once established some guidelines, we can add material as experience changes our perspec-tives and sense of values.

James A. McAleer of General Electric Company[4] notes that any state-ment of basic purchasing policy or a purchasing manual needs to be reviewed, redesigned, and overhauled at least every 3 years. Purchasing policies should not become static; nor should all construction companies try to fit into a standard mold that would effectively satisfy the needs of their operations in different locations.

A brief rundown of ideas for the new manual to progress from the informal instructions of the contractor to the more-formal listing of provisions to arrive at a working basis would include:

1. Define the purchasing agent's authority. What limitations apply, and to whom does he report for approvals or guidance?

2. Define the relationships with other functions in the company, for instance, methods of handling change estimates.

3. Develop special methods of handling subcontract bids and estimates.

4. Standardize and publicize improved practices. Develop proce-dures to eliminate departmental duplicating of correspondence, approval letters, requisitions, and orders.

5. Train new assistants in special techniques, and educate clerical help to maintain purchasing methods and improved methods of subcon-tract preparation.

6. Improve the purchasing performance, and set goals for the future.

7. Expand list of subcontractors; accumulate financial statements for later reference when needed.

8. Assemble material for purchasing checklist and manual preparation.

9. Develop unit cost files on subcontract work.

10. Keep management posted on what is being accomplished.

With respect to the last item, unless the contractor works with a small organization, there will be phases of the subcontracting operation that should be checked by a second individual or a board of directors.

After negotiations are concluded, common practice calls for the submission of a report to the management outlining the financial aspects of the subcontract and giving a comparison with the estimate, a summary of the other bids received, and special terms, if any.

The subcontractor's financial statement is ordinarily submitted together with a Dun and Bradstreet credit rating report. The net worth and working capital are two other pertinent factors in which the contractor's financial officer will be interested.

The form of the letter of intent (Fig. 10) is such that if the subcontractor's resources seem insufficient to the financial officer, a discussion can still be held regarding additional guarantees before the agreement is approved for execution by the contractor.

From the point of view of the subcontractor or salesman seeking a subcontract or an order, the average purchasing agent may appear to be relentlessly pursuing the lowest possible price for certain work or materials and haggling over the details. Granting that some buyers behave in this fashion, the experienced buyer recognizes that he probably has more contacts with subcontractors, vendors, and other companies than any other representative of the contractor. Consequently, those engaged in buying have an unusual opportunity to add to or detract from the company's reputation in the relationships with these other firms. Respect for those in purchasing presumably may be given more freely when it becomes apparent that quality and service receive reasonable consideration in any evaluation before contracting.

MOTIVATION

One task of the contractor which bears consideration in his organization is to so arrange conditions of work that the employees develop a sense of participating in the establishment of their own goals at the same time as their efforts are personally being directed toward the objectives of the company.

Writing about the techniques and concepts of motivation which apply to purchasing as well as to other phases of business life, Douglas McGregor, in *The Human Side of Enterprise,* has discussed several basic principles in human relations.[5] In substance, McGregor believes that there is a self-fulfilling prophecy in what the boss does or does not do when he makes an assumption about "people" which causes them to adjust and respond in the manner he expects.

It is interesting to note that those who are trusted tend to be motivated to expend their best efforts for producing high-quality work. Mc-Gregor's program, which lends itself to a five-step process of development, has been further explained in a pamphlet entitled "Making Management by Objective Work" by Alva F. Kindall of the Graduate School of Business Administration at Harvard University.[6]

Time for discussion and analysis, which sometimes is scarce, needs to be found for periodic sessions with his superior to size up the individual's attainment of the objective which he has set for himself. It may be that there is a desire to subscribe to a professional journal, to attend a conference, or to take an evening course for the next 6 months. He might ask for assignments which will help prepare for the next step in advancement. The key seems to be that improvement in performance can be expected as individuals can discuss their objectives and become more familiar with the company's long-term goals.

One of the primary purposes of this book is to review some of the situations that have arisen in one purchasing agent's life in the construction business as a guide primarily for the subcontracting and purchasing function, and to open up the horizon in that particular area and in the overall management of a construction company.

Whether or not the individual examples apply to the particular company or situation in which the reader is interested becomes of less importance than the comparison which he himself makes between his past experiences and the situations mentioned in the text, as they come up daily on hundreds of jobs around the country, and his ability to utilize the knowledge thus gained in discussions with his associates for mutual benefit.

Buildings will probably never be built without questions arising with respect to both the actual construction and the contract terms. The chances for controversies will always occur. Our discussion may help to avoid a few; or at least it may furnish a basis for a rational settlement that permits completion of the job.

NOTES—Chapter 9

1. Truman W. Cottom, *Contractor's Desk Book*, Prentice-Hall, Englewood Cliffs, N.J., 1965.
2. J. H. Westing, I. V. Fine, and members of the Milwaukee Association of Purchasing Agents, *Industrial Purchasing*, 3d ed., John Wiley, New York, 1955.
3. Frederick S. Merritt (ed.), *Building Construction Handbook*, McGraw-Hill, New York, 1965.
4. James A. McAleer, "Purchasing Policies and Manuals," National Association of Purchasing Managers.
5. Douglas McGregor, *The Human Side of Enterprise*, McGraw-Hill, New York, 1960.
6. Alva F. Kindall, "Making Management by Objective Work," National Association of Purchasing Managers.

CHAPTER TEN

Purchasing Department Forms and Uses

In Chapter 9 certain policies and procedures for the purchasing department of a construction company were outlined as a basis of operation. In this chapter we will develop in more detail the whys and wherefores for the special forms which have proved practical following many years' use.

With a new or smaller construction organization there is need for a continuing reappraisal of the value of procedures which may have started as a result of some individual's previous experience in purchasing work. "Is this form necessary?" and "Can we operate without it?" are two questions that should be asked frequently.

SUBCONTRACT FORMS

In Chapter 4 we touched on two types of subcontract agreement forms, i.e., a one-page version (Fig. 19) or a seven-page form (Appendix C, Form 1), which are commonly used in the construction business. The longer forms are usually developed and tempered over the years by the give and take of negotiations with subcontractors who have found them basically acceptable. The interpretations, in practice, have proved fair

AGREEMENT

AGREEMENT dated   , 19 by and between A. B. C. Construction Company, a Delaware corporation having an office at   (herein called the *"General Contractor"*), and

having an office at

                    (herein called the *"Contractor"*).

Witnesseth :

Whereas, the General Contractor has entered into an agreement with   (herein called the *"Owner"*) to construct the Building (using said term herein as the same is defined in *Section 1* of *Exhibit A* hereto) on the Site located on the

             and more particularly described in *Section 1(e)* of said *Exhibit A*; and

Whereas, the General Contractor and the Contractor have agreed that the Contractor is to undertake a portion of the work in connection with construction of the Building.

Now, Therefore, in consideration of the premises and of the mutual promises and undertakings herein contained, the parties hereto hereby agree as follows:

1. The General Contractor hereby engages the Contractor and the Contractor hereby agrees to furnish and perform the Work (using said term herein as the same is defined in *Section 1* of *Exhibit A* hereto) relating to

for the Building, all as shown upon, and in accordance with, the Drawings and Specifications identified in *Schedule 1* to *Exhibit A* hereto, including, without limitation, the furnishing and performance of all labor, materials, tools, apparatus, equipment, services, transportation, scaffolding, processes and other things required for the completion of the Contractor's duties under this Agreement, and otherwise in accordance with this Agreement, including the General Conditions and the Additional Provisions which are annexed hereto as *Exhibits A* and *B,* respectively, and which constitute parts of this Agreement.

2. Subject to the terms of this Agreement, the Contractor shall immediately proceed with the Work, and shall commence the Work at the Site

and shall perform the Work in accordance with the General Contractor's progress schedule and otherwise so as not to delay the completion of construction of the Building or any part thereof.

3. In consideration of the performance of the Work by the Contractor, the General Contractor agrees to pay to the Contractor, subject only to additions and deductions provided for in this Agreement, the sum of

the payment thereof to be made in accordance with, and subject to, the terms of this Agreement.

In Witness Whereof, the parties hereto have executed this Agreement as of the day and year first above written.

In the presence of: A. B. C. Construction Company

....   By ....
                         Vice President

In the presence of: ....
                           Contractor

....   By ....
                            Title

FIG. 19 Form of Subcontract.

and reasonable, and there is seldom any complaint from experienced subcontractors about individual provisions.

In both cases, the special provisions for the particular job are duplicated, made available in the bidding processes, and then included in the agreement. All papers should be dated, to avoid confusion over revisions, and initialed by the parties executing the agreement.

The following are indicative of some of the applicable variations in contract preparation:

Date of Agreement   Agreements executed on Sunday are generally void by statute.

Name of Contracting Parties   The exact names or legal titles of the parties should be used, including the place of business and the place of incorporation.   For example:

CORPORATION: A.B.C. Corporation, Inc., a Nebraska corporation with offices located at 3900 Broadway, Billings, Montana.

PARTNERSHIPS: John Doe and James Smith, a partnership, doing business as, or d/b/a, A.B.C. Company with offices located at 425 Bergen Street, St. Louis, Missouri.

INDIVIDUAL: John Doe, an individual, d/b/a Doe Company with offices located at 265 Hamilton Avenue, Indianapolis, Indiana.

JOINT VENTURE: A.B.C., Inc., a New York corporation, and X.Y.Z. Co., Inc., an Arizona corporation, jointly and severally, with offices located at 350 LaSalle Drive, Chicago, Illinois. (Normally, a joint venture has one office established for all business purposes.)

Signatures of Contracting Parties   Signatures must agree exactly with names of parties as first written in the agreement.   In the case of a firm, the signature of the firm name by one of the partners, in nearly all cases, binds the firm and each of its members.   It does not bind special partners, except to the extent of their interest.   It does not bind partners if the contract is for something not within the scope of the firm's business.

The best practice is to insert the name of each partner and that of the firm where names of contracting parties first appear in the agreement. If all those mentioned in the agreement as partners are in fact partners at the time of execution of the agreement, the firm's signature signed by one general partner is legally sufficient and affords ample protection.

The name of the corporation should be followed by the signature of the officer authorized to execute a contract.   The seal of the corporation should be attached or impressed and attested by the proper officer.

For a voluntary association, signatures are required of its officers and a sufficient number of responsible individual members to ensure the carrying out of the financial obligation assumed.

Authority to execute a contract

1. *Individual:* There is ordinarily no legal bar to execution by an individual of a contract for employment of an architect or for the erection of a building.

2. *Corporation:* To be established that:

    a. The corporation has the right to enter into the proposed contract.

    b. It has exercised that right by legal action.

    c. The officer executing the contract has been authorized so to act by the corporation.

3. *An authority expending public funds:* Validity of an agreement between such bodies and an architect or contractor to be determined by competent counsel.

Common practice assumes that the agreement, signed by the president, sealed with the corporate seal, and attested by the secretary, binds a corporation. A certificate should be attached to the agreement showing that the general power to sign is duly vested in the person named, or a "true copy" should be secured from the minutes attested by the secretary, with corporate seal.

Witnesses Witnesses at signing are not necessary. Witnesses may be difficult to produce in case of a contest. If there are none, signatures may be proved by any competent evidence.

Seals Attachment of a seal is a necessary part of legal execution of a contract by a corporation. A firm as such has no seal, but the attachment of a seal binds the partner executing the instrument.

The only significance of a seal in ordinary contracts is to imply a consideration. Use of a seal, therefore, except for a corporation, is unnecessary. A bond must be under seal.

Usually two copies of the subcontract form are prepared for execution, with additional conformed copies made for those requiring the information in the contractor's or architect's office.

There are occasions when the subcontractor's proposal form is offered to constitute an agreement. Generally speaking, the contractor will do well to avoid signing these forms, but should instead excerpt the technical details and incorporate them in the subcontract agreement which his purchasing man prepares. It should be recognized that proposal forms, prepared by subcontractors, may include provisions that could easily be overlooked and cause confusion later if there is the slightest variation from the architect's drawings and specifications to which the contractor will be bound.

Where there are unacceptable provisions in the proposal form and at the same time lengthy descriptions of equipment and performance which cannot be readily included in the subcontract, a clause such as the following could be used: "All materials and equipment shall comply in all respects with the physical description, specifications, performance characteristics and warranties as stated in proposal dated _____ but no other terms and conditions of the said proposal shall be part of this subcontract."

In a somewhat similar vein, some subcontract agreement forms offered by trade or professional associations while presenting a fair basis for performance also include clauses that could provoke hardship on the contractor. For instance, stipulations such as the following should be watched for: "The contractor agrees to pay the subcontractor on demand for his work or materials as far as executed and fixed in place, less the retained percentage, at the time the Certificate of Payment should issue, even though the architect fails to issue it for any cause not the fault of the subcontractor."

There are many reasons why the architect may not get around to issuing his certification even though the contractor's work is entirely satisfactory. To expect the contractor to pay his subcontractor under these conditions presents an unfair burden. It is better to agree to waive such a provision at the time of subcontracting. True, the contractor may have a cause of action against the owner or architect, but the language of the subcontract document is clear. If the work is being executed at some remote point from the main offices of the various parties, the subcontractor will be hammering at the door for payment long before the matter of issuance of the certificate is finally resolved.

SUBCONTRACT RECORD CARDS

When subcontracts are first prepared, it is customary to fill out a subcontract record card (Fig. 20), including pertinent subcontract data, with space left for change-order additions and deductions as well as insurance information. At the conclusion of the job, the cards will be filed with the permanent contract records.

RATING CARDS

Rating cards (Fig. 21) are also prepared for each new subcontractor and filed alphabetically in trade classifications. An entry of the new subcontract will be made for those who have performed subcontract work previously. Toward the conclusion of the project, a questionnaire (Fig. 22) is sent to those representatives involved in the subcontract management in each department, i.e., operations, engineering, and purchasing. The completed questionnaire is commonly circulated in the purchasing department so that the men become better acquainted with

SUBCONTRACTOR					DATE	
ADDRESS					SENT FOR SIGNATURE	
	REC'D.	EXPIRES	RENEWED	LIMITS	RETURNED	
W. C. INS. CERT						
P. L. INS. CERT					DISTRIBUTED	
P. D. INS. CERT					STATE U. I. NO.	
C. L. INS. CERT						
AUTO INS. CERT					GUARANTEE REC'D	
JOB:					FINAL PAYMENT	
WORK:						
CONT. FORM						
AMT. OF CONTRACT						
CHANGE ORDERS						

NUMBER	DATE	AMOUNT ADDED	DESCRIPTION OF CHANGES	AMOUNT DEDUCTED

FIG. 20   Subcontract Record Card.

						No.
FIRM NAME				SUBJECT		
CORPOR. OF						

DATE	JOB NAME AND NO.	AMOUNT	DELIVERY	EXECUTION	QUALITY	REMARKS
				A		
				B		
				C		
				A - START B - PROGRESS C - COMPLETION		

FIG. 21   Subcontractor Rating Card.

the subcontractor's performance and ratings on current jobs. Replies are then abstracted for entry on the permanent rating cards.

In Chapter 9 we discussed the steps in checking subcontract estimates and in following up for approval through the architect's and owner's representatives. Once this has been received, the change orders should

```
┌─────────────────────────────────────────────────────────────────────────┐
│                    A. B. C. CONSTRUCTION COMPANY          _____     │
│                      PURCHASING DEPARTMENT        Subcontract Index #      │
│                   REPORT ON SUBCONTRACTOR'S PERFORMANCE                    │
│                                                                           │
│   SUBCONTRACTOR_____  WORK_____ │
│                                                                           │
│   JOB_____LOCATION_____CONTRACT #_____S/C DATE___│
│                                                                           │
│   Office Contact_____Field Representative_____│
│                                                                           │
│   Architect_____Proj. Captain_____Engineer_____ │
│                                                             Mech./Struct. │
│   Quality Archt./Engr. dwgs. and specs. Complete ( )  Sketchy ( )         │
│                                                                           │
│   1.  ENGINEER'S REPORT                                                   │
│       a)  Time for submission - quality shop dwgs._____ │
│       b)  Conformance A.B.C. scheduling and procedures_____ │
│       c)  Submission estimates (prompt, slow)_____ │
│       d)  Pricing changes (reasonable, excessive)_____ │
│       e)  GENERAL PERFORMANCE   ┌─────────────────────────────────────┐   │
│                                 │                                     │   │
│                                 │                                     │   │
│                                 └─────────────────────────────────────┘   │
│                                      Engineer                    Date     │
│   2.  SUPERINTENDENT'S REPORT     GOOD   AVERAGE    POOR    COMMENTS       │
│       a)  Start (prompt, slow)_____ │
│       b)  Progress_____ │
│       c)  Completion_____ │
│       d)  Fabrication_____ │
│       e)  Delivery_____ │
│       f)  A.B.C. time spent expediting_____ │
│       g)  Cooperation_____ │
│       h)  Labor supply_____ │
│       i)  Labor relations_____ │
│       j)  Quality workmanship - shop_____ │
│                              field_____ │
│       k)  Punch list cleanup_____ │
│       l)  GENERAL PERFORMANCE ┌──────────────────────────────────────┐    │
│                               │                                      │    │
│       m)  Remarks_____│                                      │    │
│                               └──────────────────────────────────────┘    │
│                                      Superintendent              Date     │
│   3.  PURCHASING AGENT'S REPORT                                           │
│       a)  Type Subcontract  Lump Sum( )  Cost Plus( )  Cost Plus G.T.( )  U.P.( ) │
│       b)  New Construction ( )  Alteration ( )    Total S/C Amount $_____ │
│       c)  Business Relationship  Satisfactory ( )  Comments_____ │
│       d)  GENERAL PERFORMANCE ┌ Good ( )  Average ( )  Unsatisfactory ( ) │
│       e)  Future Work  Yes ( )  Limitations_____ │
│       f)  Comments_____ │
│                                                                           │
│   _____    _____  │
│    Project Executive        Date       Purchasing Agent            Date   │
│   COMPOSITE  ┌─────────────────────┐   _____ │
│   RATING     │                     │    Chief Purchasing Agent      Date   │
│              └─────────────────────┘                                      │
└─────────────────────────────────────────────────────────────────────────┘
```

FIG. 22   Report on Subcontractor's Performance.

be prepared, usually A-# for additions (Fig. 16) and D-# for deductions (Fig. 17). Entries are made on the subcontract record card with a revised subcontract price indicated to reflect the change. Normally, no acknowledgment is requested unless the change is of major proportions.

VOUCHER CHECKS

The voucher checks for subcontract payments are usually prepared by an accounting department representative, but the various copies of the form provide space for approvals (Appendix C, Form 2) by those responsible for approving performance: the superintendent; the accountant, who records all amounts; the purchasing agent, who verifies the accuracy of the subcontractor's account as shown on the voucher and who approves a reduction in reserve if proper at this particular time; and finally the home-office accountant, who is responsible for the accuracy of the computations and for verification that proper insurance is still in force.

Normally the project executive would be one of the individuals signing subcontract checks, with the other officer not serving actively on the particular project.

INSURANCE

Most specifications note the types and minimum limits of insurance which the contractor is required to carry. Similarly the subcontract form, completed after consultation with attorney and insurance broker, should note the applicable requirements. Generally speaking a broad form hold harmless clause offers protection for the contractor with respect to his liability arising out of operations performed by subcontractors, and is confirmed through the carrying of contractural insurance coverage by all subcontractors.

The following table indicates that the types and minimum limits of coverage may vary depending on the different degrees of hazard with respect to the subcontract work as well as on the effect due to project location. (See p. 226.)

The general liability insurance which the contractor requires is to evidence the financial ability of the subcontractor to withstand (1) suits directly against the subcontractor and (2) suits brought against the contractor in which the subcontractor is involved and which it is expected that the subcontractor will handle on the contractor's behalf by virtue of the hold harmless clause included in the agreement.

INSURANCE CERTIFICATES

Certificates are required in accordance with the limits shown in the subcontract form and/or specifications. The purchasing agent should verify the receipt of the various types, should follow for corrections if necessary, and then should issue a confirming receipt form (Fig. 23) to the superintendent showing that the subcontractor is authorized to start work on the job. Without this coverage, the subcontractor should not be authorized to start. It is necessary to keep in close touch with the expiration dates on policies which can be handled through notes on the job minutes indicating that insurance certifications have been received (insurance certificate received, expires _____).

Suggested Minimum Limits

Type of coverage	Urban site			Nonurban site		
	Per person	Two or more persons per accident	Annual aggregate	Per person	Two or more persons per accident	Annual aggregate
General liability:*						
"N" Nonhazardous:						
Bodily injury	$250,000	$1,000,000	$ 200,000	$100,000	$ 300,000	$100,000
Property damage		100,000			50,000	
"H" Hazardous:						
Bodily injury	300,000	1,500,000	500,000	250,000	1,000,000	200,000
Property damage		250,000			100,000	
"X" Extrahazardous:						
Bodily injury	300,000	1,500,000	1,000,000	250,000	1,000,000	500,000
Property damage		500,000			250,000	
Automobile liability:						
Bodily injury	250,000	1,000,000		100,000	300,000	
Property damage		100,000			50,000	

* *General Degree of Hazard:*

"N" Nonhazardous Work: Generally interior work and finishing work. Also nonconstruction trade services.

"H" Hazardous Work: Generally exterior work: structural work performed by the mechanical trades (electrical; heating, ventilating, and air conditioning; and plumbing); the iron and steel trades; and elevator trades.

"X" Extrahazardous Work: Wrecking, demolition, excavation, and foundation work. If a subcontract involves two or more types of work, rate according to the most hazardous of types involved.

A. B. C. CONSTRUCTION COMPANY
SUBCONTRACTOR'S INSURANCE CERTIFICATES

SUPT_____   OFFICE_____DATE_____

CONTRACT_____
NAME AND NUMBER

_____, *Subcontractor on the above job has furnished*
Certificates covering the following which meet the requirements of his Subcontract:

1. **Workmen's Compensation Insurance**   Expires_____
2. **Public Liability – (B. I. and P. D.)**   Expires_____
3. **Contractors' Protective Liability (B. I. and P. D.)**   Expires_____
4. **Contractual Liability (B. I. and P. D.)**   Expires_____
5. _____   Expires_____

Men should not work after expiration of insurance Certificates unless renewals have
been received and you have been so notified.
Men employed by this Subcontractor may start work at once.

COPIES TO: _____

FIG. 23   Insurance Certificate Receipt Form.

BACKCHARGES

Contractors' backcharges for work performed for subcontractors, such as billings, hoisting and handling charges, or material deliveries, are usually prepared at the field office and need to be acknowledged and processed promptly so that they may be recorded in monthly voucher statements.

From a practical point of view, when subcontractors perform work for each other, it is preferable that they bill each other rather than have the charges handled through the contractor's books. This avoids the issuance of subcontract change-order additions and deductions which would incorrectly reflect final cost records for the individual trades.

It sometimes happens that the purchasing agent has to take on the responsibility of effecting a settlement or compromise between subcontractors who may be close to litigation. If there is need for the passing of checks, there should be assurance of a sufficiency of funds and of all loose ends tied up simultaneously. Otherwise the matter may drag on for months, may prevent final settlement of the subcontractors' accounts, and may be the cause of delay in the contractor's final release and payment.

BILLS OF SALE

These forms (Appendix C, Form 3) are executed by subcontractors when payments are being made prior to incorporation of materials into the building and when a specific passing of title seems desirable.

AFFIDAVIT FORMS

These forms (Appendix C, Form 4) are executed by subcontractors upon request if a reduction in the reserve withholding is contemplated by the contractor. They may also be called for in the subcontract agreement if the subcontractor's finances are such that one feels there is need for periodic verification that his accounts are being handled currently.

The standard theories of checks and balances apply especially in arranging for final payment to subcontractors. Some companies have approval forms printed for execution by the individuals involved, with questions to be replied to in their particular province. (See Appendix C, Form 2.)

The superintendent would certify that the work is complete; that all claims for extras are processed and backcharges issued. The engineer might certify that all the subcontract guarantees and roofing bonds have been received, that the maintenance instructions on mechanical equipment have been distributed, and that the "as built" drawings have been received, etc.

The purchasing agent would confirm that all change orders have been written, that there are no known outstanding claims by others, and that the insurance had not expired as of the date of completion of the work in the field.

After these forms have been received in proper order, the final voucher can be prepared in the accounting office and a final release form (Fig. 24) sent to the subcontractor. Upon execution and return by the subcontractor, it would be understood that it would be effective only upon the receipt of a check in the stated amount of the release.

PURCHASE ORDERS

Purchase orders (Fig. 25) are commonly issued for materials or supplies needed for the job and should indicate the extended total list as well as complete delivery instructions. In some cases the phrase "delivery when requested by the superintendent" is included to avoid unexpected and too-early deliveries. The superintendent, of course, is expected to keep in close touch with his requirements and to release such orders in time to meet the job schedules.

Discount and payment terms should be shown and an acknowledgment form attached. On orders involving repeat deliveries, it is common practice to have a field copy printed to list current deliveries.

Common practice also calls for the issuance of a material receipt for the goods as delivered. When matched up with a purchase order

GENERAL RELEASE

WHEREAS, pursuant to contract made   by and between

(date of subcontract)

A. B. C. CONSTRUCTION COMPANY

and

(name of subcontractor)

hereinafter called the **SUBCONTRACTOR**, for

(brief statement of work covered by contract, name of project and its location)

final payment is about to be made,

NOW, THEREFORE, in consideration of the premises and of the sum of

Dollars **($)**

(amount of final payment)

lawful money of the United States being the full and entire sum due upon the completion of the contract aforesaid to the said Subcontractor in hand paid by A.B.C. Construction Company, receipt of which is hereby acknowledged, said Subcontractor does hereby remise, release and forever discharge A.B.C. Construction Company and

(name of owner)

of and from any and all manner of actions, suits, debts, dues, sums of money, accounts, reckonings, bonds, bills, covenants, controversies, agreements, promises, claims, demands and liens whatsoever in law or in equity which the said Subcontractor has or may have for or on account of or in connection with the contract aforesaid.

IN WITNESS WHEREOF, the Subcontractor has caused its name to be hereunto subscribed and its seal to be hereunto affixed this   day of
nineteen hundred and

ATTEST OR WITNESS:

(SEAL)

SECRETARY	SUBCONTRACTOR
ADDRESS	BY
	PRESIDENT

FIG. 24   General Release.

PURCHASE-NO.: __57__ - __3__
CONTRACT   ORDER

A.B.C. Construction Company
CLEVELAND, OHIO

DATE October 15, 1974

THIS NUMBER
MUST APPEAR
ON ALL
INVOICES
CORRESPONDENCE
AND PACKAGES

PLEASE MARK AND DELIVER

MATERIAL SPECIFIED BELOW TO ➡ { A.B.C. Construction Company
c/o A. B. Jones Building
Fremont, Ohio

ORDERED
FROM

Brown Brothers Hardware Company

Mansfield, Ohio

Mail Invoices (2) copies and Shipping Lists to
A. B. C. CONSTRUCTION CO.
c/o A. B. Jones Building
Fremont, Ohio

FOLD HERE

⌐ **The following conditions apply to this order** ¬

F.O.B. Job Site, Fremont, Ohio Delivery Dates: Superintendent's instructions
Discount Terms   Net   Sales Taxes: (X) Included; ( ) Excluded; ( ) Exempt Project
The right to cancel this order is reserved if the material specified is not shipped by date promised.

RE: A. B. Jones Building-Contract #57
Fremont, Ohio   Approval Letter A-27

FINISH HARDWARE

This will authorize you to furnish and deliver f.o.b. job
site all finish hardware required in connection with the construction of the new office building in Fremont, Ohio, for
A. B. Jones Company. All hardware shall be furnished in accordance with the drawings and specifications prepared by Smith
and Smith, architects, as listed on our project list contract
information dated July 22, 1974.

Hardware schedule is to be prepared and submitted promptly
to permit necessary coordination with hollow metal work.

All deliveries of hardware shall be made when and as directed
by our field superintendent.

For the above, we will pay you the lump sum of $2,500 plus
sales tax of $100 for a total of Two Thousand Six Hundred Dollars
($2,600).

Payment is to be made on or about the 15th day of each month
following delivery. Five percent of the value of the items delivered will be withheld until final acceptance of hardware by
A.B.C., the architect, and the owner. Final payment will be made
within 30 days after receipt of such acceptance.

A. B. C. CONSTRUCTION COMPANY

BY_____
J. J. Smith

FIELD ACCOUNTANT'S COPY
PURCHASING DEPT. FILE COPY
ACCOUNTING DEPT. COPY
SUPERINTENDENT'S COPY
OWNER'S COPY
OWNER'S COPY

**FIG. 25 Purchase Order, a multicarbon form using a different color paper for
each recipient. Only the last line of each carbon, showing the recipient, is
illustrated.**

230

copy it forms the basis for processing the invoice in the home office for payment. In some cases, the voucher checks are prepared in the field office for signature by authorized individuals.

MODIFICATIONS

Modifications of purchase orders (Fig. 26) are usually issued for revisions and major changes in quantities. Minor overfills or underfills seldom require modifications, except as they involve changes for additional or omitted work in the contract.

OTHER FILES

It is good practice to maintain files of all financial statements which have been submitted by subcontractors. Sometimes these are hard to come by, but they offer a good basis of comparison in analyzing changes in assets and net worth. Résumés or brochures are available more frequently and should be filed for reference.

Subcontractors' estimates, even though the jobs were not finalized, can prove valuable for comparison purposes, and are worth preserving for a period.

Purchasing departments normally are the most-frequent users of financial records of the type issued by Dun and Bradstreet. When unknown subcontractors are encountered in negotiation, it is good policy to request reports, a certain number of which are furnished as part of the service. It helps to confirm a subcontractor's financial picture before becoming too deeply involved. As an example, near the conclusion of some negotiation, the contractor might determine that the subcontractor completely lacks the financial strength required for a particular job.

Sales departments usually have need for registers giving names and addresses and data on corporation directors.

Engineering departments commonly maintain files of the trade catalogs for various building products as well as *Sweet's Architectural Catalog*. In all cases, the department using the data should keep the file up to date, with other departments borrowing and using on occasion.

Trade data such as the *Thomas Register of American Manufacturers* are useful for industrial subcontractors where a broader market needs to be contacted.

The Associated General Contractors of America can help in furnishing data on subcontractors' registers available in many localities in the country.

A.B.C. Construction Company

CLEVELAND, OHIO

REFER TO: __57__ - __3__
CONTRACT ORDER

DATED Oct. 15, 1974

MODIFICATION OF PURCHASE ORDER

TO

Brown Brothers Hardware Company

Mansfield, Ohio

DATE March 3, 1975

FOLD HERE

PLEASE MODIFY PURCHASE ORDER ABOVE REFERRED TO AS HEREINAFTER SPECIFIED.
RENDER INVOICES IN ACCORDANCE WITH PURCHASE ORDER AS MODIFIED.
COMPLY WITH ALL TERMS AND CONDITIONS OF PURCHASE ORDER

RE: A. B. Jones Building, Contract # 57
Fremont, Ohio

Finish Hardware

This will confirm authorization to furnish additional
hardware in accordance with your invoices noted below.

Change Estimate CE-35

Your invoice # 00812 Dated 1/22/75 $35.00
Your invoice # 00956 Dated 2/11/75   37.00

                                            $72.00

In consideration of the above there will be an addition
to the above order in the amount of Seventy-Two Dollars ($72.00).

A. B. C. CONSTRUCTION COMPANY

By _____

J. J. Smith

FIELD ACCOUNTANT'S COPY

PURCHASING DEPT. FILE COPY

ACCOUNTING DEPT. COPY

SUPERINTENDENT'S COPY

OWNER'S COPY
OWNER'S COPY

FIG. 26 Modification of Purchase Order, a multicarbon form using a different
color paper for each recipient. Only the last line of each carbon, showing the
recipient, is illustrated.

RECORD STORAGE

After a job has been closed out, final payments made, and final cost records determined, the purchasing files will normally be boxed in cardboard containers, inventoried, numbered, and stored for a period, perhaps 5 years. Many contracts have stipulations regarding storage requirements which need to be considered along with the contractor's own ideas.

There is apt to be considerable duplication of correspondence in various departments. For instance, the files of the job superintendent, accounting department, engineering department, and purchasing department presumably will all contain copies of subcontract agreements and change orders. Since individuals in the various departments are transferred or retire, a consistent company policy on which files are to be preserved and which are to be destroyed is highly desirable to make the records accessible, to serve a useful purpose, and to avoid excessive storage.

Appendix

A. Sample Supplementary Specifications

1. The contractor reserves the right to accept or reject any or all bids.
2. Bidders are expected to be experienced and familiar with the requirements and conditions imposed during the erection of generally similar constructed buildings.
3. Each bidder shall review the drawings and specifications covering the work of other trades and shall be responsible for the coordination of their work with the work of such trades. Drawings and specifications may be seen at the contractor's main office.
4. All bidders are expected to be familiar with local labor practices. No claim will be accepted for costs incurred as a result of jurisdictional or labor practices disputes.
5. Bidders are required to acquaint themselves with the contractor's subcontract form #_____, copies of which are available for inspection at the contractor's main office. The successful bidder will be required to enter into a subcontract on this form and will be obligated to conform to all the articles contained therein.
6. Subcontractor shall comply with all provisions of the subcontract form with respect to compensation and liability insurance.

   Before commencing the work, subcontractor shall procure and maintain, at its own expense, until completion and final acceptance of the work at least the following insurance from insurance companies satisfactory to the contractors:

a. Workmen's compensation and employers' liability insurance in accordance with the laws of the state in which the work is situated.

b. Contractor's public liability insurance including contractors' protective liability if the subcontractor sublets to another all or any portion of the work, with the following minimum limits:

Bodily injury, including death: $100,000 per person
$300,000 per accident

Property damage: $ 50,000 per accident
$100,000 aggregate per policy year

c. Automobile liability insurance covering all owned, non-owned and hired automobiles used in connection with the work, with the following minimum limits:

Bodily injury, including death: $100,000 per person
$300,000 per accident

Property damage: $ 50,000 per accident

Before commencing work, the subcontractor shall furnish a certificate, satisfactory to the contractor, from each insurance company showing the insurance in force, stating policy numbers, dates of expiration, and the limits of liability thereunder, and further providing that the insurance will not be canceled or changed until the expiration of at least ten (10) days after written notice of such cancellation or change has been mailed to and received by the contractor.

d. Fire insurance in an amount the subcontractor deems adequate to cover the value of his own plant, tools and equipment on the site.

e. It is the contractor's intent to implement a comprehensive safety program. Compliance with this safety program and cooperation with its administration will be a mandatory requirement of all subcontractors.

7. Bids shall include all federal, state, county, and municipal taxes imposed by law to be collected and paid by the subcontractor.

8. All permits required for the work will be obtained and paid for by the subcontractor. General construction permit will be by others.

9. Contractor will provide the following services for the benefit of subcontractors, without cost to them, except as may be otherwise stated.

a. Watchman service, but the contractor or owner will not be responsible for loss on account of theft or otherwise of the property of any subcontractor or their employees.

b. Water used for construction purposes and a water main up through the building with valved outlets and water barrel at each floor, but each subcontractor shall pay for any extensions of hose or piping which he may require; and each subcontractor shall be responsible for any damage resulting from his careless use of water.

c. Electric current for normal construction purposes consisting of 208-volt, three-phase, and 110-volt service will be available throughout the building as per contractor's specification for Temporary Light and

Power. All costs resulting from connections to this power source will be by the subcontractor. Each subcontractor will provide his own approved trailer for power equipment.

d. General temporary electric lighting as per contractor's specification for Temporary Light and Power, but each subcontractor shall pay for any extension of the lighting service which he may require.

e. Temporary toilets for workmen.

f. Temporary heat and protection on those floors on which the glazing has been completed. All temporary heating and protection for work and materials occurring on those floors not enclosed must be provided and included in the subcontractor's proposal.

g. Sidewalk bridge and fencing including walkways to the building.

h. Temporary construction personnel elevator service for the movement of men throughout the building as the progress of the work permits. Service will be provided at no cost during normal working hours plus one-half hour start-up and one-half hour shut-down time.

10. In the event the subcontractor wishes to schedule the use of the contractor's material hoist and material elevator, he will pay at the rate of $_____ per hour. This charge will include operator's time and bellman on both straight time and overtime basis.

11. Any lifts over 1,500 pounds or any lifts which would not fit in a 3 wheelbarrow, 12-foot high cage, shall be handled through other hoisting arrangements by the subcontractor at his own expense.

12. The concrete and masonry subcontractors will provide their own separate hoists, etc. Cost of electric service to these hoists will be by the subcontractor. Locations of these hoists must be approved by the contractor's superintendent. No other subcontractors will be allowed to install their own material hoists without permission of the contractor's superintendent. [This clause would normally not be included on smaller buildings where one hoist would be sufficient for all trades.]

13. Location of subcontractor's items of plant, tools, mixers, cutters, etc. must be approved by the contractor's superintendent.

14. All subcontractors shall provide their own field offices, shanties, tool sheds, trailers, etc., procuring approval of size, location, and general arrangement before setting up at the site. All subcontractors' shanties, field offices, tool sheds, etc. shall be of fireproof construction (details shall be subject to the approval of the contractor's superintendent) and shall contain adequate fire-protection equipment.

15. Subcontractor shall pay for wiring for temporary light and power for his construction shanties.

16. Should any of the subcontractor's field offices, tool sheds, shanties, material storage areas, etc. at the site of the work obstruct the progress of any portion of the work, they shall be moved by the subcontractor, without reimbursement of cost, from place to place or from the premises, as the contractor's superintendent may direct. The ground level shall be considered transient storage only, and all subcontractors must remove materials from this level within 24 hours.

17. No overhead or profit will be allowed should overtime be authorized on contract work.

18. The subcontractor shall notify the contractor's superintendent by 2:30 P.M. of any day he is working overtime so that arrangements can be made to supply the supplementary personnel.

19. Any subcontractor working beyond normal working hours will pay the added costs for providing temporary water and toilets, temporary light and power, temporary elevators, if required, unless such overtime is authorized as extra work in writing by the contractor's superintendent. This includes overtime required in order to facilitate meeting construction schedule.

20. All patching or comeback required to complete the work due to plant items of contractor or subcontractor shall be included in the subcontractor's bid.

21. Subcontractor shall include all cutting and patching required in connection with his own work.

22. Subcontractor shall be responsible for cleaning in connection with his work. Subcontractor shall remove all his own rubbish, debris, cartons, surplus materials, etc. from the building at his own expense. All materials that can be disposed of by use of the rubbish chute will be stacked at one location as directed by the contractor's superintendent. Contractor will dump this material down the chute and truck it away from the site at no cost to the subcontractor.

B. Specifications Temporary Light and Power for Construction Purposes

1. LEGAL CODE REQUIREMENTS

The electrical work for construction purposes shall conform to all federal, state, and municipal requirements. This subcontractor shall obtain and pay for any required applications, permits, and inspections pertaining to this work.

2. INSURANCE REQUIREMENTS

The requirements are the same as for the permanent installation. The work shall conform to the requirements of the "National Electrical Code."

3. GENERAL

Temporary work shall be installed in such a manner as not to interfere with the permanent construction. If such interference does occur, it will be the responsibility of the subcontractor to make such changes as may be required to overcome the interference. The cost of these changes will be included as part of the subcontract price.

4. MATERIAL

As the life of this installation is limited, and as this installation will not form a part of the finished building, minimum cost is the basic requirement, consistent with material and workmanship which will satisfactorily meet job conditions.

5. SCOPE OF WORK

Provide labor and materials for the installation and maintenance of temporary light and power as may be required during the period of construction. Cost of electricity will be by others.

a. SERVICE: Temporary three-phase, 60-cycle, 4-wire, 120–208-volts current will be brought into the building by the utility company adjacent to the main switchgear room. The electrical subcontractor will pick up from the utility company service and will distribute electrical service for light and power to all points of the building as may be required for construction purposes.

Arrangements for metering shall be as required by the utility company. The electrical subcontractor will make the installation and furnish the material required.

Service will be brought from the point of entrance through a fused safety service switch to meet the utility company requirements. A main distribution board will be furnished and installed for distribution of a total lighting load of approximately _____ kw and a power load of approximately _____ hp.

b. GENERAL LIGHTING: Distribution for general lighting will be made from the distribution board to temporary plug fuse lighting panels located on each floor. From these lighting panels, No. 12 circuit wiring with "pig tail" medium base lamp sockets will distribute lighting on the basis of an average ¼ watt per square foot for the entire floor area. Receptacles for plug-in trailers will be provided every 75 feet.

Temporary wiring will be of a movable nature and will be required at varying locations as the work of the various trades progresses. Charges for moving temporary lighting shall be included as part of the maintenance cost.

c. STAIR LIGHTING: Stair lighting is to be installed on the circuits and panels separate from the general lighting circuits. Each floor and intermediate landing is to be provided with a plain receptacle, 75-watt lamp and wire guard.

d. MISCELLANEOUS LIGHTING: Safety lights (red) are to be installed at truck entrances, fire hydrant, and fire standpipe connections. Sidewalk bridge lights for approximately _____ feet of sidewalk bridge are to be provided using 100-watt lamps and wire guards.

Temporary light for construction shanties is to be included. Furnish twenty-five 40-foot trailers, two #14 wire with attachment plug on one end and socket on the other end. Quote unit price for additional trailers.

e. POWER: The electrical subcontractor is to provide temporary power for all hoisting machines except those required for structural steel. One hoist at approximately 75 hp will be required. Three-phase motors for welding machines, floor finishing and grinding machines, pipe threaders, etc. are to be connected as required for various trades throughout the building. Wiring for this equipment must be of a movable nature and subject to change and relocation. Charges for relocations are to be included as part of the maintenance cost.

Connections for single-phase portable tools are to be made from lighting circuits and are to be made available to meet various trade requirements. Electrical service is to be provided for temporary water booster pump, and temporary ejector pumps in the basement.

Power for elevator operating on 460 volts will be supplied by means of step-up transformers. This elevator will be placed in temporary operation as promptly as possible.

f. TEMPORARY HEAT: Bidders are requested to furnish the following separate prices in connection with temporary heat. These prices are not to be included in the base price.

(1) Lump sum for maintenance of temporary heat on the basis of 135 twenty-four-hour days, seven days per week during the winter of _____.

(2) Unit price per floor for wiring two temporary unit heaters. Do not include maintenance in this price.

(3) Unit price per 8-hour overtime shift for maintenance of temporary heat.

g. SIGNAL SYSTEMS AND TELEPHONES: Furnish, install, and maintain a complete signal and telephone system for one hoist tower. Installation is to be made in accordance with the requirements of the contractor's Safety Bulletin.

h. COMMUNICATION SYSTEM: Furnish, install, and maintain a complete communication system covering all floors which will be connected to the general contractor's field office, which will be located remote from the building. System equipment shall be equal to RCA and shall consist of all necessary equipment with speakers and talk-back provision. Distribution by floors, one speaker per floor. Submit unit price for more or fewer loudspeakers.

i. LAMPS AND FUSES: Furnish and install 100-watt lamps for general circuit lighting and all fuses as may be required for a complete job. Replacement of lamps and fuses will be the responsibility of the electrical subcontractor throughout the life of the job.

6. MAINTENANCE

All temporary facilities are to be maintained and kept in good operating condition. Maintenance men necessary to perform this work shall be provided in accordance with union requirements. Maintenance time will include allowance for normal working hours for all trades including start-up and shut-down overtime as required. Regular overtime during weekdays,

Saturdays, Sundays, and holidays is not to be included. Quote unit price for overtime.

7. TRANSFER TO PERMANENT SERVICE

When power becomes available from the permanent building service, it may be desirable to transfer temporary requirements to this source. Transfers are to be made only as directed by the general contractor. Cost of such work is to be included in the electrical subcontract.

8. REMOVAL AND SALVAGE

The electrical subcontractor shall disassemble and remove from the property all temporary wiring and equipment when its use is no longer required. Surplus material will be salvaged and the salvage value reflected in the estimated price.

9. QUOTE UNIT PRICES MENTIONED ABOVE AS FOLLOWS:

a. Forty-foot trailer, #14 wire, with attachment plug and socket.
b. Labor cost per overtime shift for maintenance of electrical work on temporary heat.
c. Installed price for each added or omitted loudspeaker.
d. Labor cost per overtime hour for maintenance of temporary light and power.
e. Labor cost per straight-time shift for maintenance of electrical work on temporary heat.
f. Unit price per floor for wiring two temporary unit heaters. Do not include any maintenance cost in this price.

C. Forms

FORM 1   Subcontract.

This Agreement, made the **Fifteenth** day of **August** in the year one thousand nine hundred and **Seventy Four** by and between **A.B.C. CONSTRUCTION COMPANY,** (hereinafter called **A.B.C.**) and **Heating Company, Davies Road, Lima, Ohio, an Ohio corporation**

(hereinafter called the Subcontractor),

Witnesseth, that the Subcontractor and **A.B.C.** agree as follows:

ARTICLE I. The Subcontractor shall perform and furnish all the work, labor, services, materials, plant, equipment, tools, scaffolds, appliances and all other things necessary for **Heating, Ventilating, and Air Conditioning**

(hereinafter called the Work) for and at the **A. B. Jones Building** (hereinafter called the Project), located on premises at **Fremont, Ohio** (hereinafter called the Premises), as shown and described in and in strict accordance with the Plans, Specifications, General Conditions, Special Conditions and Addenda thereto prepared by **Smith and Smith** (hereinafter called the Architect) and with the terms and provisions of the General Contract (hereinafter called the General Contract) between A.B.C. and **the A. B. Jones Company** (hereinafter called the Owner) dated **July 1, 1974** annexed hereto and in strict accordance with the Additional Provisions, page(s) **3A** and made a part hereof.

ARTICLE II. The Plans, Specifications, General Conditions, Special Conditions, Addenda and General Contract, hereinabove mentioned, are available for examination by the Subcontractor at all reasonable times at the office of A.B.C.; all of the aforesaid, including this Agreement, being hereinafter sometimes referred to as the Contract Documents. The Subcontractor represents and agrees that it has carefully examined and understands this Agreement and the other Contract Documents, has investigated the nature, locality and site of the Work and the conditions and difficulties under which it is to be performed, and that it enters into this Agreement on the basis of its own examination, investigation and evaluation of all such matters and not in reliance upon any opinions or representations of A.B.C., or of the Owner, or of any of their respective officers, agents, servants, or employees.

With respect to the Work to be performed and furnished by the Subcontractor hereunder, the Subcontractor agrees to be bound to the Owner and to A.B.C. by each and all of the terms and provisions of the General Contract and the other Contract Documents, and to assume toward the Owner and A.B.C. all of the duties, obligations and responsibilities that A.B.C. by those Contract Documents assumes toward the Owner, and the Subcontractor agrees further that the Owner and A.B.C. shall have the same rights and remedies as against the Subcontractor as the Owner under the terms and provisions of the General Contract and the other Contract Documents has against A.B.C. with the same force and effect as though every such duty, obligation, responsibility, right or remedy were set forth herein in full. The terms and provisions of this Agreement with respect to the Work to be performed and furnished by the Subcontractor hereunder are intended to be and shall be in addition to and not in substitution for any of the terms and provisions of the General Contract and the other Contract Documents.

ARTICLE III. The Subcontractor shall complete the several portions and the whole of the Work at or before the time or times hereinafter stated, to wit:

The Subcontractor shall carefully coordinate his Work with the job requirements and shall furnish at all times sufficient material, skilled workmen, and equipment to perform the work to the entire satisfaction of A.B.C. and to conform to the schedule as it develops and so as not to delay the completion of the whole or any part of the Work.

Should the progress of the Work or of the Project be delayed by any fault or neglect or act or failure to act of the Subcontractor or any of its officers, agents, servants or employees so as to cause any additional cost, expense, liability or damage to A.B.C. or to the Owner or any damages or additional costs or expenses for which A.B.C. or the Owner may or shall become liable, the Subcontractor shall and does hereby agree to compensate A.B.C. and the Owner for and indemnify them against all such costs, expenses, damages and liability.

A.B.C., if it deems necessary, may direct the Subcontractor to work overtime and if so directed the Subcontractor shall work said overtime and, provided that the Subcontractor is not in default under any of the terms or provisions of this Agreement or of any of the other Contract Documents, A.B.C. will pay the Subcontractor for such actual additional wages paid, if any, at rates which have been approved by A.B.C., plus taxes imposed by law on such additional wages, plus workmen's compensation insurance, liability insurance and levies on such additional wages if required to be paid by the Subcontractor.

If, however, the progress of the Work or of the Project be delayed by any fault or neglect or act or failure to act of the Subcontractor or any of its officers, agents, servants or employees, then the Subcontractor shall, in addition to all of the other obligations imposed by this Agreement upon the Subcontractor in such case, and at its own cost and expense, work such overtime as may be necessary to make up for all time lost and to avoid delay in the completion of the Work and of the Project.

Article IV. The sum to be paid by A.B.C. to the Subcontractor for the satisfactory performance and completion of the Work and of all of the duties, obligations and responsibilities of the Subcontractor under this Agreement and the other Contract Documents shall be

Ninety-Nine Thousand Dollars and No Cents ($99,000.00) _____

(hereinafter called the Price) subject to additions and deductions as herein provided.

The Price includes all Federal, State, County, Municipal and other taxes imposed by law and based upon labor, services, materials, equipment or other items acquired, performed, furnished or used for or in connection with the Work, including but not limited to sales, use and personal property taxes payable by or levied or assessed against the Owner, A.B.C. or the Subcontractor. Where the law requires any such taxes to be stated and charged separately, the total price of all items included in the Work plus the amount of such taxes shall not exceed the Price.

On or before the last day of each month the Subcontractor shall submit to A.B.C., in the form required by A.B.C., a written requisition for payment showing the proportionate value of the Work installed to that date, from which shall be deducted: a reserve of ten per cent (10%); all previous payments; and all charges for services, materials, equipment and other items furnished by A.B.C. to or chargeable to the Subcontractor; and the balance of the amount of such requisition, as approved by A.B.C. and the Architect and for which payment has been received by A.B.C. from the Owner, shall be due and paid to the Subcontractor on or about the fifteenth (15th) day of the succeeding month.

The Subcontractor shall submit with its first requisition for payment a detailed schedule showing the breakdown of the Price into its various parts for use only as a basis of checking the Subcontractor's monthly requisitions.

A.B.C. reserves the right to advance the date of any payment (including the final payment) under this Agreement if, in its sole judgment, it becomes desirable to do so.

The Subcontractor agrees that, if and when requested so to do by A.B.C., it shall furnish such information, evidence and substantiation as A.B.C. may require with respect to the nature and extent of all obligations incurred by the Subcontractor for or in connection with the Work, all payments made by the Subcontractor thereon, and the amounts remaining unpaid, to whom and the reasons therefor.

The final payment shall be due within forty (40) days after completion and acceptance of the Work by A.B.C. and the Architect, provided first, however, that (1) A.B.C. shall have received final payment therefor from the Owner, (2) the Subcontractor shall have furnished evidence satisfactory to A.B.C. that there are no claims, obligations or liens outstanding or unsatisfied for labor, services, materials, equipment, taxes or other items performed, furnished or incurred for or in connection with the Work and (3) the Subcontractor shall have executed and delivered in a form satisfactory to A.B.C. a General Release running to and in favor of A.B.C. and the Owner. Should there prove to be any such claim, obligation or lien after final payment is made, the Subcontractor shall refund to A.B.C. all monies that A.B.C. and/or the Owner shall pay in satisfying, discharging or defending against any such claim, obligation or lien or any action brought or judgment recovered thereon and all costs and expenses, including legal fees and disbursements, incurred in connection therewith.

If any claim or lien is made or filed with or against A.B.C., the Owner, the Project or the Premises by any person claiming that the Subcontractor or any subcontractor or other person under it has failed to make payment for any labor, services, materials, equipment, taxes or other items or obligations furnished or incurred for or in connection with the Work, or if at any time there shall be evidence of such nonpayment or of any claim or lien for which, if established, A.B.C. or the Owner might become liable and which is chargeable to the Subcontractor, or if the Subcontractor or any subcontractor or other person under it causes damage to the Work or to any other work on the Project, or if the Subcontractor fails to perform or is otherwise in default under any of the terms or provisions of this Agreement, A.B.C. shall have the right to retain from any payment then due or thereafter to become due an amount which it deems sufficient to (1) satisfy, discharge and/or defend against any such claim or lien or any action which may be brought or judgment which may be recovered thereon, (2) make good any such nonpayment, damage, failure or default, and (3) compensate A.B.C. and the Owner for and indemnify them against any and all losses, liability, damages, costs and expenses, including legal fees and disbursements, which may be sustained or incurred by either or both of them in connection therewith. A.B.C. shall have the right to apply and charge against the Subcontractor so much of the amount retained as may be required for the foregoing purposes.

— 2 —

If the amount retained is insufficient therefor, the Subcontractor shall be liable for the difference and pay the same to A.B.C.

No payment (final or otherwise) made under or in connection with this Agreement shall be conclusive evidence of the performance of the Work or of this Agreement, in whole or in part, and no such payment shall be construed to be an acceptance of defective, faulty or improper work or materials nor shall it release the Subcontractor from any of its obligations under this Agreement; nor shall entrance and use by the Owner constitute acceptance of the Work or any part thereof.

ARTICLE V. Should the Subcontractor be delayed in the commencement, prosecution or completion of the Work by the act, omission, neglect or default of A.B.C. or of anyone employed by A.B.C., or of any other contractor or subcontractor on the Project, or by any damage caused by fire or other casualty or by the combined action of workmen in no wise chargeable to the Subcontractor, or by any extraordinary conditions arising out of war or government regulations, or by any other cause beyond the Subcontractor's control and not due to any fault, neglect, act or omission on its part, then the Subcontractor shall be entitled to an extension of time only, such extension to be for a period equivalent to the time lost by reason of any and all of the aforesaid causes, as determined by A.B.C. In the event of dispute by the Subcontractor, the matter shall be referred to the Architect whose decision thereon shall be final and binding upon the parties hereto. The Subcontractor shall not be entitled to any such extension of time, however, unless a claim therefor is presented in writing to A.B.C. within forty-eight (48) hours of the commencement of such claimed delay. Such extension or extensions of time, as determined by A.B.C. or by the Architect, or the decision or decisions of the Architect that no extension of time shall be allowed, shall release and discharge A.B.C. of and from any and all claims of whatever character by the Subcontractor on account of the aforesaid or any other causes of delay.

ARTICLE VI. The Subcontractor in making or ordering shipments shall not consign or have consigned materials, equipment or any other items in the name of A.B.C. A.B.C. is under no obligation to make payment for charges on shipments made by or to the Subcontractor but may, at its option, pay such charges, in which case the Subcontractor shall reimburse A.B.C. for the amount of such payments plus a service charge of twenty-five per cent (25%) of the amount so paid.

ARTICLE VII. Notwithstanding the dimensions given on the Plans, Specifications and other Contract Documents it shall be the obligation and responsibility of the Subcontractor to take such measurements as will insure the proper matching and fitting of the Work covered by this Agreement with contiguous work.

The Subcontractor shall prepare and submit to A.B.C. such shop drawings as may be necessary to describe completely the details and construction of the Work. Approval of such shop drawings by A.B.C. and/or the Architect shall not relieve the Subcontractor of its obligation to perform the Work in strict accordance with the Plans, Specifications, the Additional Provisions hereof and the other Contract Documents, nor of its responsibility for the proper matching and fitting of the Work with contiguous work.

Should the proper and accurate performance of the Work hereunder depend upon the proper and accurate performance of other work not covered by this Agreement, the Subcontractor shall carefully examine such other work, determine whether it is in fit, ready and suitable condition for the proper and accurate performance of the Work hereunder, use all means necessary to discover any defects in such other work, and before proceeding with the Work hereunder, report promptly any such improper conditions and defects to A.B.C., in writing and allow A.B.C. a reasonable time to have such improper conditions and defects remedied.

ARTICLE VIII. The Work hereunder is to be performed and furnished under the direction and to the satisfaction of both the Architect and A.B.C. The decision of the Architect as to the true construction, meaning and intent of the Plans and Specifications shall be final and binding upon the parties hereto. A.B.C. will furnish to the Subcontractor such additional information and Plans as may be prepared by the Architect to further describe the Work to be performed and furnished by the Subcontractor and the Subcontractor shall conform to and abide by the same.

The Subcontractor shall not make any changes, additions and/or omissions in the Work except upon written order of A.B.C. as provided in Article IX hereof.

ARTICLE IX. A.B.C. reserves the right, from time to time, whether the Work or any part thereof shall or shall not have been completed, to make changes, additions and/or omissions in the Work as it may deem necessary, upon written order to the Subcontractor. The value of the work to be changed, added or omitted shall be stated in said written order and shall be added to or deducted from the Price.

The value of the work to be changed, added or omitted shall be determined by the lump sum or unit prices, if any, stipulated herein for such work. If no such prices are stipulated, such value shall be determined by whichever of the following methods or combination thereof A.B.C. may elect:

(a) By adding or deducting a lump sum or an amount determined by a unit price agreed upon between the parties hereto.

(b) By adding (1) the actual net cost to the Subcontractor of labor in accordance with the established rates, including required union benefits, premiums the Subcontractor is required to pay for workmen's compensation and liability insurance, and payroll taxes on such labor, (2) the actual cost to the Subcontractor of materials and equipment and such other direct costs as may be approved by A.B.C., less all savings, discounts, rebates and credits, (3) an allowance of 5% for overhead on items (1) and (2) above, and (4) an allowance of 5% for profit on items (1), (2) and (3) above.

Should the parties hereto be unable to agree as to the value of the work to be changed, added or omitted, the Subcontractor shall proceed with the work promptly under the written order of A.B.C. from which order the stated value of the work shall be omitted, and the determination of the value of the work shall be referred to the Architect whose decision shall be final and binding upon the parties hereto.

In the case of omitted work A.B.C. shall have the right to withhold from payments due or to become due to the Subcontractor an amount which, in A.B.C.'s opinion, is equal to the value of such work until such time as the value thereof is determined by agreement or by the Architect as hereinabove provided.

— 3 —

FORM 1   (Continued).

Additional Provisions

    All work shall be performed in accordance with drawings A 1 to
A 10 Inclusive, E 1 to 5 Inclusive, HVAC 1 to 5 Inclusive, P 1-2
Inclusive, and Specifications all dated July 2, 1974, subject to the
following provisions:
    (1) All work described as Alternate #1 in the bidding information
covering Temporary Heat is included in the scope of this subcontract.
    (2) The furnishing of concrete equipment bases is excluded from
this subcontract but the subcontractor will supply sketches, templates
and anchor bolts as required for his work and shall cooperate with
others in the proper and accurate installation of such anchorage.

3A

All changes, additions or omissions in the Work ordered in writing by A.B.C. shall be deemed to be a part of the Work hereunder and shall be performed and furnished in strict accordance with all of the terms and provisions of this Agreement and the other Contract Documents.

ARTICLE X. The Subcontractor shall at all times provide sufficient, safe and proper facilities for the inspection of the Work by A.B.C., the Architect and their authorized representatives in the field, at shops, or at any other place where materials or equipment for the Work are in the course of preparation, manufacture, treatment or storage. The Subcontractor shall, within twenty-four (24) hours after receiving written notice from A.B.C. to that effect, proceed to take down all portions of the Work and remove from the premises all materials whether worked or unworked, which the Architect or A.B.C. shall condemn as unsound, defective or improper or as in any way failing to conform to this Agreement or the Plans, Specifications or other Contract Documents, and the Subcontractor, at its own cost and expense, shall replace the same with proper and satisfactory work and materials and make good all work damaged or destroyed by or as a result of such unsound, defective, improper or nonconforming work or materials or by the taking down, removal or replacement thereof.

ARTICLE XI. Should the Subcontractor at any time refuse or neglect to supply a sufficiency of skilled workmen or materials of the proper quality and quantity, or fail in any respect to prosecute the Work with promptness and diligence, or cause by any act or omission the stoppage or delay of or interference with or damage to the work of A.B.C. or of any other contractors or subcontractors on the Project, or fail in the performance of any of the terms and provisions of this Agreement or of the other Contract Documents, or should the Architect determine that the Work or any portion thereof is not being performed in accordance with the Contract Documents, or should there be filed by or against the Subcontractor a petition in bankruptcy or for an arrangement or reorganization, or should the Subcontractor become insolvent or be adjudicated a bankrupt or go into liquidation or dissolution, either voluntarily or involuntarily or under a court order, or make a general assignment for the benefit of creditors, or otherwise acknowledge insolvency, then in any of such events, each of which shall constitute a default hereunder on the Subcontractor's part, A.B.C. shall have the right, in addition to any other rights and remedies provided by this Agreement and the other Contract Documents or by law, after three (3) days written notice to the Subcontractor mailed or delivered to the last known address of the latter, (a) to perform and furnish through itself or through others any such labor or materials for the Work and to deduct the cost thereof from any monies due or to become due to the Subcontractor under this Agreement, and/or (b) to terminate the employment of the Subcontractor for all or any portion of the Work, enter upon the premises and take possession, for the purpose of completing the Work, of all materials, equipment, scaffolds, tools, appliances and other items thereon, all of which the Subcontractor hereby transfers, assigns and sets over to A.B.C. for such purpose, and to employ any person or persons to complete the Work and provide all the labor, services, materials, equipment and other items required therefor. In case of such termination of the employment of the Subcontractor, the Subcontractor shall not be entitled to receive any further payment under this Agreement until the Work shall be wholly completed to the satisfaction of A.B.C. and the Architect and shall have been accepted by them, at which time, if the unpaid balance of the amount to be paid under this Agreement shall exceed the cost and expense incurred by A.B.C. in completing the Work, such excess shall be paid by A.B.C. to the Subcontractor; but if such cost and expense shall exceed such unpaid balance, then the Subcontractor shall pay the difference to A.B.C. Such cost and expense shall include, not only the cost of completing the Work to the satisfaction of A.B.C. and the Architect and of performing and furnishing all labor, services, materials, equipment, and other items required therefor, but also all losses, damages, costs and expenses, including legal fees and disbursements sustained, incurred or suffered by reason of or resulting from the Subcontractor's default.

ARTICLE XII. A.B.C. shall not be responsible for any loss or damage to the Work to be performed and furnished under this Agreement, however caused, until after final acceptance thereof by A.B.C. and the Architect, nor shall A.B.C. be responsible for loss of or damage to materials, tools, equipment, appliances or other personal property owned, rented or used by the Subcontractor or anyone employed by it in the performance of the Work, however caused.

A.B.C. or the Owner shall effect and maintain fire insurance (with extended coverage, if specified or otherwise required) upon all Work, materials and equipment incorporated in the Project and all materials and equipment on or about the Premises intended for permanent use or incorporation in the Project or incident to the construction thereof, the capital value of which is included in the cost of the Work, but not including any contractors' machinery, tools, equipment, appliances or other personal property owned, rented or used by the Subcontractor or anyone employed by it in the performance of the Work.

The total value of the property described above as insurable under this Article and as shown on the approved monthly requisition provided for in Article IV, plus the total value of similar property incorporated in the Project or delivered on the Premises during the month but not included in said requisition, as reported by the Subcontractor to A.B.C. for insurance purposes only, shall determine the total value of the Subcontractor's work, materials and equipment to be insured under this Article.

The maximum liability to the Subcontractor under this insurance shall be for not more than that proportion of any loss which the last reported value of the insured property bore to the actual value of said property at the time of such last report, and in no event for more than the actual loss.

In the event of a loss insured under this Article, the Subcontractor shall be bound by any adjustment which shall be made between A.B.C. or the Owner and the insurance company or companies. Loss, if any, shall be made payable to A.B.C. and/or the Owner, as their interests may appear, for the account of whom it may concern.

ARTICLE XIII. The subcontractor shall, at its own cost and expense, (1) keep the Premises free at all times from all waste materials, packaging materials and other rubbish accumulated in connection with the execution of its Work by collecting and depositing said materials and rubbish in locations or containers as designated by A.B.C. from which it shall be removed by A.B.C. from the Premises without charge, (2) clean and remove from its own Work and from all contiguous work of others any soiling, staining, mortar, plaster, concrete or dirt caused by the

— 4 —

execution of its Work and make good all defects resulting therefrom, (3) at the completion of its Work in each area, perform such cleaning as may be required to leave the area "broom clean," and (4) at the entire completion of its Work, remove all of its tools, equipment, scaffolds, shanties and surplus materials. Should the Subcontractor fail to perform any of the foregoing to A.B.C.'s satisfaction, A.B.C. shall have the right to perform and complete such work itself or through others and charge the cost thereof to the Subcontractor.

ARTICLE XIV. The Subcontractor shall obtain and pay for all necessary permits and licenses pertaining to the Work and shall comply with all Federal, State, Municipal and local laws, ordinances, rules, regulations, orders, notices and requirements, including, among others, those relating to discrimination in employment, fair employment practices or equal employment opportunity, and with the requirements of the Board of Fire Underwriters, whether or not provided for by the Plans, Specifications, General Conditions or other Contract Documents, without additional charge or expense to A.B.C., and shall also be responsible for and correct, at its own cost and expense, any violations thereof resulting from or in connection with the performance of its Work. The Subcontractor shall at any time upon demand furnish such proof as A.B.C. may require showing such compliance and the correction of such violations. The Subcontractor agrees to save harmless and indemnify A.B.C. from and against any and all loss, injury, claims, actions, proceedings, liability, damages, fines, penalties, costs and expenses, including legal fees and disbursements, caused or occasioned directly or indirectly by the Subcontractor's failure to comply with any of said laws, ordinances, rules, regulations, orders, notices or requirements or to correct such violations.

ARTICLE XV. The Subcontractor shall not employ men, means, materials or equipment which may cause strikes, work stoppages or any disturbances by workmen employed by the Subcontractor, A.B.C. or other contractors or subcontractors on or in connection with the Work or the Project or the location thereof. The Subcontractor agrees that all disputes as to jurisdiction of trades shall be adjusted in accordance with any plan for the settlement of jurisdictional disputes which may be in effect either nationally or in the locality in which the Work is being done and that it shall be bound and abide by all such adjustments and settlements of jurisdictional disputes, provided that the provisions of this Article shall not be in violation of or in conflict with any provisions of law applicable to the settlement of such disputes. Should the Subcontractor fail to carry out or comply with any of the foregoing provisions, A.B.C. shall have the right, in addition to any other rights and remedies provided by this Agreement or the other Contract Documents or by law, after three (3) days written notice mailed or delivered to the last known address of the Subcontractor, to terminate this Agreement or any part thereof or the employment of the Subcontractor for all or any portion of the Work, and, for the purpose of completing the Work, to enter upon the Premises and take possession, in the same manner, to the same extent and upon the same terms and conditions as set forth in Article XI of this Agreement.

ARTICLE XVI. The Subcontractor for the Price herein provided for, hereby accepts and assumes exclusive liability for and shall indemnify, protect and save harmless A.B.C. and the Owner from and against the payment of :

1. All contributions, taxes or premiums (including interest and penalties thereon) which may be payable under the Unemployment Insurance Law of any State, the Federal Social Security Act, Federal, State, County and/or Municipal Tax Withholding Laws, or any other law, measured upon the payroll of or required to be withheld from employees, by whomsoever employed, engaged in the Work to be performed and furnished under this Agreement.

2. All sales, use, personal property and other taxes (including interest and penalties thereon) required by any Federal, State, County, Municipal or other law to be paid or collected by the Subcontractor or any of its subcontractors or vendors or any other person or persons acting for, through or under it or any of them, by reason of the performance of the Work or the acquisition, ownership, furnishing or use of any materials, equipment, supplies, labor, services or other items for or in connection with the Work.

3. All pension, welfare, vacation, annuity and other union benefit contributions payable under or in connection with labor agreements with respect to all persons, by whomsoever employed, engaged in the Work to be performed and furnished under this Agreement.

ARTICLE XVII. The Subcontractor hereby agrees to indemnify, protect and save harmless A.B.C. and the Owner from and against any and all liability, loss or damage and to reimburse A.B.C. and the Owner for any expenses, including legal fees and disbursements, to which A.B.C. and the Owner may be put because of claims or litigation on account of infringement or alleged infringement of any letters patent or patent rights by reason of the Work or materials, equipment or other items used by the Subcontractor in its performance.

ARTICLE XVIII. The Subcontractor for itself and for its subcontractors, laborers and materialmen and all others directly or indirectly acting for, through or under it or any of them covenants and agrees that no mechanics' liens or claims will be filed or maintained against the Project or Premises or any part thereof or any interests therein or any improvements thereon, or against any monies due or to become due from the Owner to A.B.C. or from A.B.C. to the Subcontractor, for or on account of any work, labor, services, materials, equipment or other items performed or furnished for or in connection with the Work, and the Subcontractor for itself and its subcontractors, laborers and materialmen and all others above mentioned does hereby expressly waive, release and relinquish all rights to file or maintain such liens and claims and agrees further that this waiver of the right to file or maintain mechanics' liens and claims shall be an independent covenant and shall apply as well to work, labor and services performed and materials, equipment and other items furnished under any change order or supplemental agreement for extra or additional work in connection with the Project as to the original Work covered by this Agreement.

If any subcontractor, laborer or materialman of the Subcontractor or any other person directly or indirectly acting for, through or under it or any of them files or maintains a mechanic's lien or claim against the Project or Premises or any part thereof or any interests therein or any improvements thereon or against any monies due or to become due from the Owner to A.B.C. or from A.B.C. to the Subcontractor, for or on account of any work, labor, services, materials, equipment or other items performed or furnished for or in connection with the Work or under any change order or supplemental agreement for extra or additional work in connection with the Project, the Subcontractor agrees to cause such liens and claims to be satisfied, removed or discharged at its own expense by bond, payment or otherwise within ten (10) days from the date of the filing thereof, and upon its failure so to

— 5 —

do A.B.C. shall have the right, in addition to all other rights and remedies provided under this Agreement and the other Contract Documents or by law, to cause such liens or claims to be satisfied, removed or discharged by whatever means A.B.C. chooses, at the entire cost and expense of the Subcontractor (such cost and expense to include legal fees and disbursements). The Subcontractor agrees to indemnify, protect and save harmless A.B.C. and the Owner from and against any and all such liens and claims and actions brought or judgments rendered thereon, and from and against any and all loss, damages, liability, costs and expenses, including legal fees and disbursements, which A.B.C. and/or the Owner may sustain or incur in connection therewith.

ARTICLE XIX. Neither this Agreement nor any monies due or to become due hereunder shall be assignable without the prior written consent of A.B.C. nor shall the whole or any part of this Agreement be sublet without like prior written consent. Any such assignment or subletting without such prior written consent shall be void and of no effect and shall vest no right or right of action in the assignee or subcontractor against A.B.C. A.B.C.'s consent to any assignment or subletting shall not relieve the Subcontractor of any of its agreements, duties, responsibilities or obligations under this Agreement and the other Contract Documents, and the Subcontractor shall be and remain as fully responsible and liable for the defaults, neglects, acts and omissions of its assignees and subcontractors and all persons directly or indirectly employed by them as it is for its own defaults, neglects, acts and omissions and those of its own officers, agents, servants and employees. The Subcontractor shall bind each of its subcontractors to all of the terms, provisions and covenants of this Agreement and the other Contract Documents with respect to the sublet Work. A.B.C.'s consent to any subletting shall not be deemed to create any contractual relationship between A.B.C. and any subcontractor to whom the Work or any portion thereof is sublet, and shall not vest any right or right of action in such subcontractor against A.B.C.

ARTICLE XX. A.B.C. shall have the right at any time by written notice to the Subcontractor, to terminate this Agreement and require the Subcontractor to cease work hereunder, in which case, provided the Subcontractor be not then in default, A.B.C. shall indemnify the Subcontractor against any damage directly resulting from such termination, except that the Subcontractor shall not be entitled to anticipated profits on work unperformed or on materials or equipment unfurnished.

ARTICLE XXI. The Subcontractor hereby guarantees the Work to the full extent provided in the Plans, Specifications, General Conditions, Special Conditions and other Contract Documents.

The Subcontractor shall remove, replace and/or repair at its own expense and at the convenience of the Owner any faulty, defective or improper work, materials or equipment discovered within one (1) year from the date of the acceptance of the Project as a whole by the Architect and the Owner or for such longer period as may be provided in the Plans, Specifications, General Conditions, Special Conditions or other Contract Documents.

Without limitation by the foregoing, the Subcontractor shall pay in addition for all damage to the Project resulting from defects in the Work and all costs and expenses necessary to correct, remove, replace and/or repair the Work and any other work or property which may be damaged in correcting, removing, replacing or repairing the Work.

ARTICLE XXII. The Subcontractor agrees that the prevention of accidents to workmen engaged upon or in the vicinity of the Work is its responsibility. The Subcontractor agrees to comply with all laws, ordinances, rules, regulations, codes, orders, notices and requirements concerning safety as shall be applicable to the Work and with the safety standards established during the progress of the Work by A.B.C. When so ordered, the Subcontractor shall stop any part of the Work which A.B.C. deems unsafe until corrective measures satisfactory to A.B.C. have been taken, and the Subcontractor agrees that it shall not have nor make any claim for damages growing out of such stoppages. Should the Subcontractor neglect to take such corrective measures, A.B.C. may do so at the cost and expense of the Subcontractor and may deduct the cost thereof from any payments due or to become due to the Subcontractor. Failure on the part of A.B.C. to stop unsafe practices shall in no way relieve the Subcontractor of its responsibility therefor.

ARTICLE XXIII. The Subcontractor hereby assumes entire responsibility and liability for any and all damage or injury of any kind or nature whatever (including death resulting therefrom) to all persons, whether employees of the Subcontractor or otherwise, and to all property caused by, resulting from, arising out of or occurring in connection with the execution of the Work; and if any claims for such damage or injury (including death resulting therefrom) be made or asserted, whether or not such claims are based upon A.B.C.'s alleged active or passive negligence or participation in the wrong or upon any alleged breach of any statutory duty or obligation on the part of A.B.C., the Subcontractor agrees to indemnify and save harmless A.B.C., its officers, agents, servants and employees from and against any and all such claims, and further from and against any and all loss, cost, expense, liability, damage or injury, including legal fees and disbursements, that A.B.C., its officers, agents, servants or employees may directly or indirectly sustain, suffer or incur as a result thereof and the Subcontractor agrees to and does hereby assume, on behalf of A.B.C., its officers, agents, servants and employees, the defense of any action at law or in equity which may be brought against A.B.C., its officers, agents, servants or employees upon or by reason of such claims and to pay on behalf of A.B.C., its officers, agents, servants and employees, upon its demand, the amount of any judgment that may be entered against A.B.C., its officers, agents, servants or employees in any such action. In the event that any such claims, loss, cost, expense, liability, damage or injury arise or are made, asserted or threatened against A.B.C., its officers, agents, servants or employees, A.B.C. shall have the right to withhold from any payments due or to become due to the Subcontractor an amount sufficient in its judgment to protect and indemnify it and its officers, agents, servants and employees from and against any and all such claims, loss, cost, expense, liability, damage or injury, including legal fees and disbursements, or A.B.C., in its discretion, may require the Subcontractor to furnish a surety bond satisfactory to A.B.C. guaranteeing such protection, which bond shall be furnished by the Subcontractor within five (5) days after written demand has been made therefor.

Before commencing the Work, the Subcontractor shall procure and maintain, at its own expense, until completion and final acceptance of the Work at least the following insurance from insurance companies satisfactory to A.B.C.:

1. WORKMEN'S COMPENSATION AND EMPLOYERS' LIABILITY INSURANCE in accordance with the laws of the State in which the Work is situated.

— 6 —

2. CONTRACTORS' PUBLIC LIABILITY INSURANCE INCLUDING CONTRACTUAL LIABIL-
ITY INSURANCE AGAINST THE LIABILITY ASSUMED HEREINABOVE, and including
CONTRACTORS' PROTECTIVE LIABILITY INSURANCE if the Subcontractor sublets to another
all or any portion of the Work, with the following minimum limits:

Bodily Injury, including death: $ 100,000   per person

$ 300,000   per accident

Property damage: $ 50,000   per accident

$ 100,000   aggregate per policy year

3. AUTOMOBILE LIABILITY INSURANCE covering all owned, non-owned and hired automobiles used
in connection with the Work, with the following minimum limits:

Bodily Injury, including death: $ 100,000   per person

$ 300,000   per accident

Property damage: $ 50,000   per accident

Before commencing the Work, the Subcontractor shall furnish a certificate, satisfactory to **A.B.C.**, from each
insurance company showing that the above insurance is in force, stating policy numbers, dates of expiration, and
limits of liability thereunder, and further providing that the insurance will not be cancelled or changed until the
expiration of at least ten (10) days after written notice of such cancellation or change has been mailed to and
received by **A.B.C.**

If the Subcontractor fails to procure and maintain such insurance, **A.B.C.** shall have the right to procure
and maintain the said insurance for and in the name of the Subcontractor and the Subcontractor shall pay the cost
thereof and shall furnish all necessary information to make effective and maintain such insurance.

ARTICLE XXIV. The Subcontractor shall furnish to **A.B.C.** a performance bond in the amount of $ ---
and a separate payment bond in the amount of $ ---   , the form and contents
of such bonds and the Surety or Sureties thereon to be satisfactory to **A.B.C.**

ARTICLE XXV. This Agreement constitutes the entire agreement between the parties hereto. No oral
representations or other agreements have been made by **A.B.C.** except as stated in this Agreement. This Agreement
may not be changed in any way except as herein provided, and no term or provision hereof may be waived by
A.B.C. except in writing signed by its duly authorized officer or agent. The marginal descriptions of any term or
provision of this Agreement are for convenience only and shall not be deemed to limit, restrict or alter the content,
meaning or effect thereof.

The said parties, for themselves, their heirs, executors, administrators, successors and assigns, do hereby
agree to the full performance of all of the terms and provisions herein contained.

In Witness Whereof the parties to these presents have hereunto set their hands as of the day
and year first above written.

In the Presence of:

A.B.C. CONSTRUCTION COMPANY

By:_____
                                        Vice-President

Heating Company

_____
                                        Subcontractor

In the Presence of:

_____

By:_____

_____
                                        Official Title

Subcontractor's_____State Unemployment Ins. No. _____
(Insert State and Register No. for State in which the Work is to be performed)

Subcontractor's License No. _____
(Insert License No., if any, for State or locality in which the Work is to be performed)

— 7 —

248

FORM 2 Departmental Approval Form. A different color is used for each page for easy identification.

A. B. C. CONSTRUCTION COMPANY

OFFICE   **DATE**

GENERAL SUPERINTENDENT ---
SUPERINTENDENT ---
PURCHASING AGENT ---
ENGINEER ---

CONTRACT:

SUBCONTRACT:

FINAL SUBCONTRACT ESTIMATE HAS BEEN RECEIVED COVERING FINAL PAYMENT TO THE ABOVE NAMED SUBCONTRACTOR. PLEASE ANSWER THE FOLLOWING QUESTIONS AND RETURN THIS FORM PROMPTLY TO THE ACCOUNTING DEPARTMENT - SUBCONTRACT DIVISION.

---

1. DO YOU KNOW OF ANY CLAIMS AGAINST THIS SUBCONTRACTOR BY THE OWNER OR BY ANY OTHER SUBCONTRACTOR?_____

2. HAVE YOU PERSONALLY INSPECTED THE WORK?_____

3. IF YOU HAVE, IS IT ENTIRELY SATISFACTORY?_____

4. HAS A REPRESENTATIVE OF THE OWNER OR ARCHITECT INSPECTED THE WORK?_____

5. DID THE WORK MEET WITH THE APPROVAL OF SUCH REPRESENTATIVE?_____

6. DO YOU KNOW OF ANY CLAIMS BY THIS SUBCONTRACTOR THAT HAVE NOT BEEN SETTLED?_____

7. HAVE WE ANY CLAIMS AGAINST THIS SUBCONTRACTOR THAT HAVE NOT BEEN SETTLED?_____

8. HAS THE PURCHASING AGENT COVERED BY PROPER CHANGE ORDER ANY SETTLEMENT THAT MAY HAVE BEEN AUTHORIZED BY YOU?_____

9. IS THERE ANY REASON WHY FINAL PAYMENT SHOULD NOT BE MADE?_____

10. HAS THERE BEEN ANY INTIMATION OF ANY KIND TO THE EFFECT THAT THIS SUBCONTRACTOR HAS NOT PAID MATERIAL BILLS OR OTHER CHARGES ENTERING INTO THE WORK?_____

(IF ANSWERS NEED FURTHER EXPLANATION, USE REVERSE SIDE.)

SIGNED_____

DATE_____

GENERAL SUPERINTENDENT

FORM 2 (Continued).

A. B. C. CONSTRUCTION COMPANY

OFFICE **DATE**

GENERAL SUPERINTENDENT ---
SUPERINTENDENT ---
PURCHASING AGENT ---
ENGINEER ---

CONTRACT:

SUBCONTRACT:

FINAL SUBCONTRACT ESTIMATE HAS BEEN RECEIVED COVERING FINAL PAYMENT TO THE ABOVE NAMED SUBCONTRACTOR. PLEASE ANSWER THE FOLLOWING QUESTIONS AND RETURN THIS FORM PROMPTLY TO THE ACCOUNTING DEPARTMENT - SUBCONTRACT DIVISION.

1. WHAT WAS THE LAST DAILY WORK INVOICE NUMBER ISSUED BY THE JOB COVERING CHARGES AGAINST ANY SUBCONTRACTOR? NO._____

2. HAVE ALL CHARGES AGAINST THIS SUBCONTRACTOR BEEN REPORTED ON DAILY WORK INVOICE BEARING A PREVIOUS NUMBER TO THE ABOVE?_____

3. DO YOU KNOW OF ANY CLAIMS AGAINST THIS SUBCONTRACTOR BY THE OWNER OR BY ANY OTHER SUBCONTRACTOR?_____

4. HAVE YOU PERSONALLY INSPECTED THE WORK?_____

5. IF YOU HAVE, IS IT ENTIRELY SATISFACTORY?_____

6. HAS A REPRESENTATIVE OF THE OWNER OR ARCHITECT INSPECTED THE WORK?_____

7. DID THE WORK MEET WITH THE APPROVAL OF SUCH REPRESENTATIVE?_____

8. DO YOU KNOW OF ANY CLAIMS BY THIS SUBCONTRACTOR THAT HAVE NOT BEEN SETTLED?_____

9. HAVE WE ANY CLAIMS AGAINST THIS SUBCONTRACTOR THAT HAVE NOT BEEN SETTLED?_____

10. HAS THE PURCHASING AGENT COVERED BY CHANGE ORDER ANY SETTLEMENT THAT MAY HAVE BEEN AUTHORIZED BY YOU?_____

11. IS THERE ANY REASON WHY FINAL PAYMENT SHOULD NOT BE MADE?_____

12. HAS THERE BEEN ANY INTIMATION OF ANY KIND THAT THIS SUBCONTRACTOR HAS NOT PAID MATERIAL BILLS OR OTHER CHARGES ENTERING INTO THE WORK?_____

13. FINAL PAYMENT IS DUE_____DAYS AFTER COMPLETION. WHEN WAS THE WORK ACTUALLY COMPLETED?_____

(IF ANSWERS NEED FURTHER EXPLANATION, USE REVERSE SIDE.)

SIGNED_____

DATE_____

SUPERINTENDENT

FORM 2 (Continued).

A. B. C. CONSTRUCTION COMPANY

OFFICE   DATE

GENERAL SUPERINTENDENT ---
SUPERINTENDENT ---
PURCHASING AGENT ---
ENGINEER ---

CONTRACT:

SUBCONTRACT:

FINAL SUBCONTRACT ESTIMATE HAS BEEN RECEIVED COVERING FINAL PAYMENT TO THE ABOVE NAMED SUBCONTRACTOR. PLEASE ANSWER THE FOLLOWING QUESTIONS AND RETURN THIS FORM PROMPTLY TO THE ACCOUNTING DEPARTMENT - SUBCONTRACT DIVISION.

1. DO YOU KNOW OF ANY OBJECTIONS OF THE OWNER OR ARCHITECT TO THE WORK COVERED BY THIS SUBCONTRACT?_____

2. HAVE ALL CERTIFICATES OF APPROVAL BEEN RECEIVED FROM CITY OR STATE AUTHORITIES HAVING JURISDICTION OVER THE WORK COVERED BY THIS SUBCONTRACT?_____

3. IS THERE ANY REASON WHY THIS FINAL PAYMENT SHOULD NOT BE MADE?_____
    IF THERE IS, PLEASE STATE REASON:

4. HAS THERE BEEN ANY INTIMATION OF ANY KIND TO THE EFFECT THAT THIS SUBCONTRACTOR HAS NOT PAID MATERIAL BILLS OR OTHER CHARGES ENTERING INTO THE WORK?_____

5. HAVE ALL WRITTEN GUARANTEES, WAIVERS OF LIEN, OR OTHER DOCUMENTS REQUIRED BY THE SPECIFICATIONS BEEN FURNISHED BY THE SUBCONTRACTOR?_____

(IF ANSWERS NEED FURTHER EXPLANATION, USE REVERSE SIDE.)

SIGNED_____

DATE_____

ENGINEER

FORM 2   (Continued).

A. B. C. CONSTRUCTION COMPANY

OFFICE   **DATE**

GENERAL SUPERINTENDENT ---
SUPERINTENDENT ---
PURCHASING AGENT ---
ENGINEER ---

CONTRACT:

SUBCONTRACT:

FINAL SUBCONTRACT ESTIMATE HAS BEEN RECEIVED COVERING FINAL PAYMENT TO THE ABOVE NAMED SUBCONTRACTOR. PLEASE ANSWER THE FOLLOWING QUESTIONS AND RETURN THIS FORM PROMPTLY TO THE ACCOUNTING DEPARTMENT - SUBCONTRACT DIVISION.

---

1.   WHAT IS THE NUMBER OF THE LAST CHANGE ORDER?_____

2.   WHAT IS THE TOTAL AMOUNT OF THIS CONTRACT ACCORDING TO YOUR RECORDS?_____

3.   DO YOU KNOW OF ANY CLAIMS BY THIS SUBCONTRACTOR THAT HAVE NOT BEEN SETTLED?_____

4.   HAVE WE ANY CLAIMS AGAINST THIS SUBCONTRACTOR THAT HAVE NOT BEEN SETTLED?_____

5.   HAVE YOU MADE ANY SETTLEMENT THAT HAS NOT BEEN COVERED BY PROPER CHANGE ORDER?_____

6.   DO YOU KNOW OF ANY OBJECTIONS OF THE OWNER OR ARCHITECT TO THE WORK COVERED BY THIS SUBCONTRACT?_____

7.   IS THERE ANY REASON WHY THIS FINAL PAYMENT SHOULD NOT BE MADE?_____

8.   HAS THERE BEEN ANY INTIMATION OF ANY KIND TO THE EFFECT THAT THIS SUBCONTRACTOR HAS NOT PAID MATERIAL BILLS OR OTHER CHARGES ENTERING INTO THE WORK?_____

9.   HAVE ALL WRITTEN GUARANTEES, WAIVERS OF LIEN, OR OTHER DOCUMENTS REQUIRED BY THE SPECIFICATIONS BEEN FURNISHED BY THE SUBCONTRACTOR?_____

(IF ANSWERS NEED FURTHER EXPLANATION, USE REVERSE SIDE.)

SIGNED_____

DATE_____

PURCHASING AGENT

FORM 3   Bill of Sale.

BILL OF SALE TO A.B.C. CONSTRUCTION COMPANY

From

_____

KNOW ALL MEN BY THESE PRESENTS

THAT,

(hereinafter called the Seller), for and in consideration of the sum of One Dollar ($1.00) and other valuable considerations to it in hand paid by A. B. C. CONSTRUCTION COMPANY, a Delaware corporation with its principal office at [   ] (hereinafter called the Purchaser), receipt whereof is hereby acknowledged, does hereby sell, transfer and assign to the said Purchaser the goods and chattels described and located on the premises described in Schedule A listed on page 4 hereto, manufactured or in process of manufacture for delivery on the

And in the event that any of said goods and chattels are in process of manufacture, it is expressly intended to sell, transfer and convey the same in their completed state, as well as in the form and state they possess at the time of the execution of this instrument.

TO HAVE AND TO HOLD all and singular the said goods and chattels to the said Purchaser, its successors and assigns to their own use and behoof forever.

And the said Seller does hereby covenant with the said Purchaser that it is the lawful owner of said goods and chattels; that they are free from all liens and claims whatsoever; that it has good right to sell the same; that it will warrant and defend same against the claims and demands of all persons.

The undersigned will provide safe and proper storage for the said goods and chattels on its own premises or other premises as may be described in the said Schedule until such a time as said goods and chattels are delivered to the said job.

The undersigned will cause to be placed conspicuously and securely on the said described goods and chattels in its plant or warehouse or other place mentioned in said Schedule, as the case may be, a sign or signs which will show that the said described property is the property of the A. B. C. CONSTRUCTION COMPANY.

The undersigned will cause the said described property to be insured against fire, theft and all other hazards in at least the value of the goods and chattels described in Schedule A and the undersigned does hereby agree to indemnify and hold the Purchaser harmless by reason of the payment made to the undersigned for any loss, theft or destruction of the said described property or any part thereof, notwithstanding the payment made to the undersigned or the transfer of title to the property in this bill of sale. The policies of insurance provided for in this paragraph shall be for the benefit of A. B. C. CONSTRUCTION COMPANY and shall be in form satisfactory to it, and the amount of such insurance shall be increased from time to time in proportion to the increase in value of said goods by any further process of manufacture.

The obligation for the performance of the contract between the undersigned and A. B. C. CONSTRUCTION COMPANY shall continue in full force and effect including but not limited to the obligation to deliver the said property pursuant to the terms of said contract.

IN WITNESS WHEREOF the said Seller has caused this instrument to be duly executed and signed this

_____day of_____, 19_____.

_____

Seller

By:_____

Official Title

FORM 3 (Continued).

FOR CORPORATION

STATE OF_____)

: SS:

COUNTY OF_____)

On the_____day of_____in the year One Thousand Nine Hundred and

_____before me personally came_____,

to me known, who being by me duly sworn, did depose and say:

That he resides in_____;

that he is _____

(President or other office)

of the _____ Corporation

(Name of Corporation)

described in and which executed the above instrument; that he knows the seal of said corporation; that the seal affixed to said instrument is such corporate seal, and that it was so affixed by order of the Board of Directors of said corporation, and that he signed his name thereto by like order.

_____

NOTARY PUBLIC

FOR INDIVIDUAL

STATE OF_____)

: SS:

COUNTY OF_____)

On the_____day of_____in the year One Thousand Nine Hundred and

_____before me personally came_____,

known to me, and to me known to be the person described in and who executed the foregoing instrument, and he acknowledged that he executed the same.

_____

NOTARY PUBLIC

FORM 3   (Continued).

SCHEDULE A

The full description of the material covered hereunder, including sizes, number, etc., together with place of location and part of premises where this material is stored, is as follows:

FORM 4 Affidavit.

STATE OF .... }
COUNTY OF .... } SS:

...., being duly sworn, deposes and says:
(Individual)

1. That he is the .... of ....
   (Title)   (Company)

Subcontractor of A. B. C. CONSTRUCTION COMPANY for ....
                                                                    (Type of Work)

at ....
     (Job Name and Location of Work)

2. That all of the statements herein contained are true and correct and that he makes this affidavit in order to induce A. B. C. CONSTRUCTION COMPANY to make payment to the said Subcontractor with full knowledge and intent that A. B. C. CONSTRUCTION COMPANY will rely thereon.

3. That the names and addresses of all persons who have not been paid in full for work, labor, services or materials performed or furnished to or for the said Subcontractor upon or in connection with the prosecution of its subcontract work, and the amounts due or to become due to such persons for such work, labor, services or materials are as follows:

NAME	ADDRESS	AMOUNT DUE OR TO BECOME DUE

4. That all Federal and State Income and other Taxes withheld or required to be withheld from the wages and salaries of all persons employed by the said Subcontractor upon or in connection with its subcontract work have been set aside and have been or will be paid when due, and that there are no such taxes presently due and unpaid or payable, except as follows:

DESCRIPTION OF TAX	AMOUNT

FORM 4 (Continued).

5. That all employer contributions required to be paid by said Subcontractor under union labor agreements with respect to persons employed by it on or in connection with its subcontract work have been or will be paid when due, and that there are no such contributions presently due and unpaid or payable, except as follows:

DESCRIPTION OF UNION CONTRIBUTION   TO WHOM PAYABLE   AMOUNT

6. That all other taxes, unemployment insurance and other contributions required to be paid by the said Subcontractor by reason of its employment of persons on or in connection with its subcontract work have been or will be paid when due, and that there are no such taxes, unemployment insurance and other contributions presently due and unpaid or payable, except as follows:

DESCRIPTION OF TAX OR CONTRIBUTION   AMOUNT

7. That all payments or advances received from A.B.C. CONSTRUCTION COMPANY by said Subcontractor for or in connection with its subcontract work will be held as trust funds to be applied first to the payment in full of the foregoing obligations before using the same for any other purpose.

(signed) ....

....
(Title)

Subscribed and sworn to before me a Notary Public this ....

day of ...., 19............

....
Notary Public

Bibliography

Books

Aljian, G. W.: *Purchasing Handbook*, 3d ed., McGraw-Hill, New York, 1973.

Bush, Vincent G: *Handbook for the Construction Superintendent*, Reston Publishing Company, Inc., Reston, Va., 1973.

Cottom, Truman W.: *Contractor's Desk Book*, Prentice-Hall, Englewood Cliffs, N.J., 1965.

Crimmins, Robert, Reuben Samuels, and Bernard Monahan: *Construction Rock-Work Guide* (Practical Construction Guide Series), John Wiley, New York, 1972.

Foxhall, W. B.: *Professional Construction Management and Project Administration*, Architectural Record, 1972.

Leonards, G. A. (ed.): *Foundation Engineering*, McGraw-Hill, New York, 1962.

Merritt, F. S. (ed.): *Building Construction Handbook*, 2d ed., McGraw-Hill, New York, 1965.

O'Brien, J. J.: *CPM in Construction Management: Project Management with CPM*, 2d ed., McGraw-Hill, New York, 1971.

Oppenheimer, S. P.: *Directing Construction for Profit: Business Aspects of Contracting*, McGraw-Hill, New York, 1971.

Park, W. R.: *The Strategy of Contracting for Profit*, Prentice-Hall, Englewood Cliffs, N.J., 1966.

Prentis, Edmund Astley, and Lazuras White: *Underpinning: Its Practice and Applications*, 2d rev. & enl. ed., Columbia University Press, New York, 1950. Includes excellent glossary of terms used in underpinning.

Sumichrast, M., and S. A. Frankel: *Profile of the Builder and His Industry*, National Association of Home Builders, 1970.

Walker, N., and Rohdenburg, T. K.: *Legal Pitfalls in Architecture, Engineering, and Building Construction*, McGraw-Hill, New York, 1968.

Westing, J. H., I. V. Fine, and members of the Milwaukee Association of Purchasing Agents: *Industrial Purchasing*, 3d ed., John Wiley, New York, 1955.

Publications of Robert Morris Associates,
Philadelphia National Bank Building,
Philadelphia, Pa. 19107

Bristow, F. E.: "Specialty Contractors' Financing Program."

Cerny, R. G.: "Construction Industry Finances—Musical Chairs with Money," 1969.

Kiddoo, G. C.: "Loans to Contractors," 1952.

Sparling, P.: "A Look at Contractor Loans from the Credit Standpoint."

"Banker-Surety Relationship: An Impasse?"

"Are You Really Informed on Contractor Loans?"

"Statement Studies," 1973.

Publications of The Associated General
Contractors of America, Inc., 1957 E Street, N.W.,
Washington, D.C. 20006

"Building Estimate Summary," #16, n.d.

"Construction—A Man's Work," n.d.

"Contractors Experience Questionnaire and Financial Statement," 1965.

"Prequalification Statement—Form #40," 1965.

"Processing Change Orders and Disputes on Federal Construction Contracts," n.d.

"Suggested Guide and Check List for Subcontracts," 1959.

"#8800 Job Overhead Summary," n.d.

Miscellaneous Publications

American Arbitration Association: "Delays and Disputes in Building Construction," 140 West 51st Street, New York 10020, 1970.

Internal Revenue Service: *Business Income Tax Returns, 1969*, Publication #453, December 1971.

Sokol, A. J., Jr.: "Contractor or Manipulator?" *Credit and Financial Management*, February 1969.

Standard Form of Agreement between Contractor and Subcontractor, Document AIA A-401 American Institute of Architects, 1735 New York Avenue, N.W. Washington, D.C. 20006.

Information from Banking Sources

The First National Bank of Chicago

James A. Bourke, "Financing Problems in the Construction Industry," address at National Electrical Contractors meeting, Atlanta, Ga., June 1971.

James A. Bourke, "Lending to Contractors," address at National Electrical Contractors Meeting, Atlanta, Ga.

Bank of America, San Francisco, Calif. 94120

"Small Business Report-Building Contractors Understanding Financial Statements"

Miscellaneous Guides

Audit of Construction Contractors, American Institute of Certified Public Accountants, 666 Fifth Avenue, New York 10019, 1965.

Profit Analyzer for Suppliers to the Building Trades, Dun and Bradstreet, Inc., 99 Church Street, New York 10008

Uniform Construction Index, The Construction Specifications Institute, Washington, D.C., 1972.

Index